Agile 2

For Mary –
Thanks for the
great ideas!
Enjoy –
Lisa Cooney

Agile 2

The Next Iteration of Agile

Cliff Berg
Kurt Cagle
Lisa Cooney
Philippa Fewell
Adrian Lander
Raj Nagappan
Murray Robinson

WILEY

This book is dedicated to the members of the global Agile 2 team, who shared their expertise and experience and collaborated remotely throughout the COVID-19 pandemic to create Agile 2.

—Cliff

I dedicate this to those who have felt frustrated that Agile, for all its hype and promise, didn't seem to bring to you much in the way of advantage and frequently seemed an exercise in pointlessness. There is goodness in Agile methods, but Agile is a hammer, and sometimes you need a violin.

—Kurt

This book is dedicated to Agilists everywhere, those struggling in the trenches to apply its values and principles, regardless of their role or field of work, and who inspired us to try to make it better.

—Lisa

This is dedicated to fellow Agilists seeking better balance, understanding, and improvement over what's been in practice for the last 20 years.

— Philippa

My contribution here is dedicated to independent thinking and independent thinkers, who prefer to choose discovering what is really out there over acquired bias or simply following marketed experts without healthy challenge and validation and who have the courage to stand up even against popular opinion if that makes sense and do not shun putting in considerable effort to develop their (independent thinking) capabilities. In a first year of a series of global pandemic challenges, my heart is with those affected and who are fighting without complaining, never giving up, and finding ways to still help others.

—Adrian

This book is dedicated to on-the-ground workers everywhere who have worked with and struggled with Agile in the past, and who inspired us to try to make it better. And to my late beagle Bindi, whose boundless love and loyalty taught me the true meaning of servant leadership.

—Raj

This book is dedicated to people who have become disenchanted with the dogma, commercialization, and fake Agile that has taken over the Agile community. We want to reclaim Agile and revive the community's ability to learn and adjust and tackle problems that the original manifesto authors did not imagine. We hope you will join us.

—Murray

About the Authors

Cliff Berg is a consultant and founder of Agile Griffin, which specializes in merging Agile and DevOps. Cliff began his career doing systems analysis for electronic systems design and then building compilers, was on the team that created the VHDL language, and wrote the first "synthesis" compiler for that language. In 1995 he cofounded and served as CTO of Digital Focus, a startup that grew to 200 people by 2000 and adopted Agile (eXtreme Programming) in full force that year. Digital Focus was sold in 2006, and since then Cliff has helped more than ten organizations adopt Agile and DevOps methods, working with leadership and teams to implement change. Cliff has experience with Agile and DevOps in a wide range of contexts, from large multiproduct digital platforms to embedded systems.

Kurt Cagle is the community editor of Data Science Central (Tech Target) and the editor in chief of the Cagle Report. He is the author of 22 books on internet technology, data modeling, and knowledge representation, and he has served as an invited expert to the W3C for more than 10 years. As a consultant, Kurt's clients have included Fortune 50 companies, and US and European government agencies. When not trying to keep a handle on what's happening in the data world, he writes novels. He can be found on LinkedIn.

Lisa Cooney currently serves as the Agile coach at Axios. She is the editor of *Evolvagility: Growing an Agile Leadership Culture from the Inside Out*, by Michael Hamman (2019). Her eclectic background includes a master's degree in education, years of being a stay-at-home mom, and substitute teaching in K-12 schools while raising children, creating art, and writing. A lifelong learner, Lisa went from Basic and Pascal college classes to creating computer-based-training at Kodak to creating her own website in HTML to virtual online learning design. She spent years designing, writing, and managing (with Agile!) the program and project management curriculum (which included systems thinking) at the Department of Homeland Security. In 2014, she designed and wrote two virtual instructor-led courses called Agile for the Federal Government and Agile for the Product Owner for the Department of Veterans Affairs.

Lisa helps organize the DC Women in Agile (WIA) meetup, is on the program committee for the Business Agility Institute's conferences in NYC, and speaks at conferences about cognitive bias and developmental feedback. She is certified as an Agile Certified Practitioner (PMI-ACP), as a ScrumMaster (CSM), and as both an Agile team facilitator (ICP-ATF) and an Agile Coach (ICP-ACC). She is working on her certification for coaching DevOps. Lisa earned her master's degree in Instructional Design from the University of Virginia and holds a bachelor's degree from Wellesley College. She can be found on LinkedIn.

Philippa Fewell has had a long-tenured, progressive consulting career leading to her current position as the Managing director of Agile Services for CC Pace Systems. Philippa has more than 35 years of experience managing and delivering complex financial, healthcare, and technology projects for both Fortune 500 companies and startups. She is an accomplished executive and hands-on manager, with demonstrated expertise in driving and supporting executive mandates to facilitate and enable business agility. Philippa has an exceptional track record of managing stakeholders at all organizational levels, with emphasis on establishing cultures of trust and building high-performance teams.

Philippa has practiced Agile methods with development teams for the last 15 years, has worked extensively with leadership on what it really means to be Agile, is a certified Agile coach, and regularly speaks on the topic of Agile benefits and practices. Philippa is highly recognized for her in-depth knowledge and practical experience in Agile, bringing hands-on, real-world techniques to the teams and the executives she works with. Philippa earned a BS in computer science with a concentration in business from Rensselaer Polytechnic Institute. She can be found on LinkedIn.

Adrian Lander has since 1995 had leadership and senior advisory roles for more than 30 well-known (Fortune 500 and FTSE 100) international client organizations, across most industries and governments. These include more than 20 Agile (and DevOps) transformations, plus other business transformations, often involving innovation and the latest technology.

Adrian began as a self-taught software developer at a young age, and before studying and researching artificial intelligence and natural language processing at several universities, he was already doing software sales and project management services for clients.

He worked for a decade as a top-level management consultant for a global consultancy, where he specialized in rescuing large, traditional programs, some beyond $100 million USD, by changing delivery to an Agile approach. In 2007, he moved to Asia to pioneer Agile in various countries, including division- and organization-wide transformations in locales including Singapore and Hong Kong, where as the lead coach, he won business awards with his teams. The past decade he has been increasingly involved in turning "bad Agile" into better Agile that has had significant measurable business benefits.

In addition to 25+ years of Agile experience, he has extensive experience in applying his expertise in the domains of professional coaching of executives, leadership and (other) teams, organizational change management, product management, and program management.

Over the years, he has obtained more than 20 professional certifications, diplomas, and degrees across domains relevant to Agile. For Adrian, however, only outcomes matter, and he believes more in the value of his skills and experience gained through deep, broad, and challenging practice. He is a founder of Agnostic Agile, a movement supported by 2,500+ Agile practitioners, focusing on openness, inclusion, and ethics in the practice of Agile. He can be found on LinkedIn.

Raj Nagappan is the founder and CEO of Catum—a software startup in the product management area. Prior to founding Catum, Raj worked as a lead engineer and manager for Nuix, which made innovative evidence discovery software for the legal industry and police. He helped it grow from a small startup to more than 500 employees with customers in more than 70 countries. Before that, he was a senior engineer in diverse organizations from startups to software vendors to investment and retail banks.

Raj holds a PhD in computer science from the Australian National University and a First Class Honours in science (majoring in computer science) from the University of Sydney. He is a Professional Scrum Master and Professional Scrum Product Owner, both from Scrum.org.

Throughout his more than 20 years of experience, Raj has sought to achieve collaboration between product/design/engineering and sales and marketing to gain a deeper understanding of customer needs and thus to craft a better "whole of product" experience. Being an engineer himself and working directly with other engineers, designers, product managers, and business stakeholders, Raj has grown frustrated with the frequent shortcomings of conventional Agile implementations and the failure of the industry to address these concerns. From 2018 he started writing about these problems and suggesting possible ways to overcome them. This naturally led to his involvement in Agile 2. He can be found on LinkedIn.

Murray Robinson works with organizations to design digital initiatives and organizations capable of realizing them. He has 30 years' experience in IT starting as a software developer, including 20 in product and project management, 16 with Agile, and 4 in UX. He has delivered large digital programs of more than $20 million with distributed teams of up to 100 people for large corporations. He is an experienced Agile coach, trainer, and practitioner with experience leading a successful Agile transformation that turned around a struggling digital agency. As an adaptive leader, he is known for getting things done and bringing enthusiasm, insight, and humor to every engagement.

Murray has an MBA from Melbourne Business School and has certifications of Professional Scrum Master from Scrum.Org, Leading SAFE 4.5 from Scaled Agile and Business Agility, and Agile Fundamentals and Agile Product Owner certifications from ICAgile. He is an ICAgile Authorized Instructor who teaches Agile, product ownership, coaching, and design thinking. He speaks about Agile and design thinking at industry events and writes about Agile on LinkedIn and his blog at agileinsights.wordpress.com. He can be found on LinkedIn.

Acknowledgments

We would like to thank the rest of the Agile 2 team—those who contributed to Agile 2's development but who are not authors of this book. They are Huet Landry, Lakshmi Chirravuri, MC Moore, Navneet Nair, Parul Choudhary, Priya Mayilsamy, Vigneshwaran Kennady, and Vincent Harris.

We would like to thank those who reviewed the draft Agile 2 content early, prior to its release, including David Anderson, Alidad Hamidi, Maarten Dalmjin, and Shane Hastie.

We would like to thank Navneet Nair for his insights in Chapter 7, "It's All About the Product."

We would like to thank Lakshmi Chirravuri for her contributions regarding PeopleOps in Chapter 14, "Agile 2 in Service Domains."

We would like to thank some individuals who provided early insights and feedback about Agile 2 ideas, including Chris Mills, Ebony Nicole Brown, Neil Green, and Thomas Fuller.

We thank the original group of Agilists who, throughout the 1990s and early 2000s, shared their wisdom and experience so generously and openly with the world.

We thank the DevOps community, including early authors such as Jez Humble and Gene Kim, for connecting the dots about continuous delivery and DevOps and explaining it to the world.

Finally, we would like to thank all those among the software and product development communities who have demonstrated original thinking in trying to improve the ecosystem and who have helped clients and colleagues to consider their actual situation instead of following a standard template. It was those who inspired true understanding of what is needed to be agile and identified gaps in current practices that need to be remedied. This includes those who were courageous enough to speak or write about dysfunctions within the community, ultimately contributing to the ideas that are behind Agile 2.

Contents

Foreword

You can recognize a great book by its ability to make obvious what is wrong with existing worldviews and to add new insights or nuances to change or improve this worldview. I believe *Agile 2* meets these criteria easily. It is an easy read not only for those new to the subject of agility, but also for die-hard professionals who are looking for something "beyond" the basic Agile concepts that are at times dogmatic and often being misused in practice.

In this era of digital disruption and an ever-growing world full of volatility, uncertainty, complexity, and ambiguity, it becomes increasingly important for organizations to become more agile. The Agile Manifesto, originally intended in 2001 to disrupt traditional, not-too-effective software development practices, has inspired many organizations over the past decades to change their ways of working, affecting both work cultures and structures. Its popularity was boosted by a growing workforce consisting of millennials and Generation-Z professionals who demand more autonomy, ownership, and the opportunity to make meaningful impact. Over the past decade, the "Agile movement" has gained increasing momentum and also moved beyond the realm of IT.

A side effect of its success and growth was that all kinds of Agile frameworks, doctrines, and certifications popped up to standardize and monetize the discipline. The original Agile values and principles, being high-level on purpose, gave ample room for various interpretations of the core paradigms. During my career I have worked with many different Agile coaches and consultants, and I was always surprised by how much discussion and fanatic debates arose among them with regard to how to live certain Agile values or implement specific practices. This tribalism led to confusion among non-Agilists, and this hampers Agile transformations significantly. I thus see a clear need for a comprehensive Agile idea set that is both pragmatic and nuanced by nature. Enter *Agile 2*.

I was happy to find out quickly that the authors do not claim to have written yet another Agile doctrine or "Bible." Instead, they have written a pragmatic companion guide that will be useful for managers and specialists alike. It is packed with hands-on tactics and practices that can help leaders and specialists in organizations to grow to a next level of agility, while preventing cargo-cult behavior or *avant le lettre* implementations that often do more harm than good.

One of my key drivers for cofounding the DevOps Agile Skills Association (DASA) in 2016 was building a comprehensive view on how to create high-performance IT organizations. The popularity of DASA stems largely from the six DevOps and Agile principles that advocate continuous improvement, customer centricity, autonomous multidisciplinary teams, and product thinking. Following these principles often results in a digital and organizational transformation that typically goes far beyond choosing a standard Agile framework or adding some basic Agile rituals to the mix. To transform successfully to high-performance, organizations need a more mature take on and guidance on what it means to really "be Agile" at scale. Providing this guidance is one of this book's core differentiating features.

Over the past decade I learned firsthand as a consultant, trainer, and senior leader the importance of building the right type of leadership in the organization and creating a culture of continuous learning, experimentation, and innovation. What I like about *Agile 2* is that both the importance of leadership and learning are advocated strongly. It provides many tangible ideas to reimagine an organization's leadership culture. I wish that I had this book on my nightstand five years ago. It would have helped me greatly in understanding why certain things happened—or did not happen—during the organizational transformations I was leading.

I like the fact that the authors do not intend to reinvent the wheel, but are keen on building on what is already working. Some of the key Agile values and principles are powerful to this day, but application in practice often needs some additional clarity and lots of examples. The authors nicely provide nuance to how to interpret Agile principles and values while referring to many interesting, and more recent, bodies of work. The authors hit the nail with addressing key topics that are haunting many organizations, leaders, and teams, such as how to collaborate, communicate, value both experts and generalists, and commit team capacity. They rightfully argue that how to adopt certain principles or how to interpret certain values depends on your organization's needs and its current level of maturity. Using this book as your Agile guide, you can aim and navigate your transformation in a more tailor-made way, resulting in more business value. I expect this book to be found on many nightstands in the coming years.

Dr. Rik Farenhorst
Senior IT Exec | Trainer | Coach | Speaker |
Writer on Creating High-Performance Digital Organizations | Co-founder of DevOps Agile Skills Association (DASA)
Utrecht, The Netherlands
December 2020

Preface

A few people who have become aware of Agile 2 have dismissed it as "more of that Agile stuff," not realizing that Agile 2 is a departure from the original Agile in attitude, approach, and substance. One of those individuals—a chemical engineer—said that he had discussed Agile at length with an Agile advocate, but still concluded that Agile is not for him. Another—an experienced systems engineer who has testified before Congress regarding aircraft and spacecraft systems reliability—also believes that Agile methods do not provide a robust process for trustworthy systems.

We view ourselves as Agilists, and yet we find widespread doubt about the efficacy and usefulness of Agile in many quarters. One of these is among engineers. These people are not ignorant. They know their job extremely well, yet Agile, as described to them, or as they have experienced it, has not resonated or has not answered critical questions.

The Agile movement also uprooted the product design community to some degree (which we will document in this book), although this is an area in which the Agile community has realized the issue and some are trying to rectify it.

Agile authors largely ignored the role of data: something that is so immensely important, that it is akin to speaking about mountains but missing a vast canyon immediately beside you.

The Agile community also sidestepped the issue of leadership — something that the DevOps community has tried to address. Leadership is so important for any endeavor, that to omit it is, frankly, quite equivocal.

Agile has not resonated among the growing DevOps community. Even though DevOps ideas were developed by people who strongly identified as Agilists, the Agile community at large has remained mostly ignorant of DevOps, which had the effect that DevOps became its own movement. As a result, most Agile coaches today know little about DevOps, and we find that DevOps practitioners often view Agile as superfluous.

You might think that mainstream programmers accept Agile, since they are the ones who use it most directly, but in actuality, there is a lot of doubt about Agile within programming communities in general. That is the biggest irony of all: that Agile, which was created for programmers, has in effect been taken away from them, and no longer serves them.

Agile is mostly accepted within *Agile* communities—comprised of Agile coaches, and managers who have been persuaded of the benefits of Agile. Programmers tend to have mixed feelings about Agile. (We will support that assertion in this book.)

Was Agile described poorly? Is Agile missing things? Did it get some things wrong? Does Agile truly not apply to the needs of the work of any of these people? Since Agile ideas can be applied to most things (in our opinion), we believe that the last explanation is not likely to be the true one.

What we have observed ourselves is that too often, Agilists explain and advocate Agile ideas and methods before asking enough questions. Some of us have seen Agile coaches fired for coming into a setting and insisting on particular practices before actually understanding how the work in that setting is done—a hypocrisy given that Agile coaches so often explain that Agile transformation is a learning journey.

To understand how to apply Agile ideas, one must first understand that domain, how the work is currently done, and why it is done that way. No Agile practice is universal. One size does not fit all, so prescribing before understanding is potentially destructive.

Indeed, the "dogma" of the Agile community helped to launch it, but it has also been its chief failing. Early proponents of Agile insisted that the Agile movement needed to be disruptive—a "call to arms"—and so dogma was called for; but dogmatic insistence also alienates and causes dysfunction when it is not the best advice for the situation.

Agile 2 is not dogmatic. It is not designed to stir up emotion. It is not a call to arms or an attempt to disrupt what we have. As such, it does not try to be disruptive. It does not replace Agile or replace DevOps or replace anything. Agile 2 pivots Agile in some important ways and attempts to fine-tune it. Agile 2 also adds many extremely crucial ideas that have been ignored by much of the Agile community, even though successful Agile practitioners often use those very ideas, and other communities of thought embrace those ideas.

Agile 2 reinforces some DevOps ideas, and some Lean ideas, but Agile 2 does not attempt to duplicate or replace those, and so those sets of ideas are still important in their own right. Agile 2 does not attempt to subsume any existing community of thought. Agile 2 also does not claim to cover all aspects of these topics. Agile 2 claims to only be a set of useful ideas for how to achieve agility in human endeavors and encourages people to include other ideas and fields of thought as well.

Agile 2 is more verbose than the Agile Manifesto. The reason is that we feel that one of the weaknesses of the manifesto was that it over-simplified complex issues. A simple value maxim cannot describe important trade-offs, and one or two principles cannot address nuanced issues such as what good leadership looks like and which styles of leadership apply best in a given situation. So Agile 2 gives these important topics the space they deserve, from an agility perspective.

One important way that Agile 2 departs from the Agile Manifesto is that Agile 2 provides the foundation of thought from which it was derived. Rather than make bold statements without substantiating them, Agile 2 provides the "problems" and "insights" that arose in discussions about Agile, in the course of the Agile 2 team's retrospective about the state of Agile. It was these problems and insights that led to the Agile 2 principles.

An Agile 2 principle is not intended to be an absolute. This is because there can be no absolutes when it comes to human behavior. An Agile 2 principle is a proposed rule of thumb: true most of the time, but perhaps not in some circumstances. That is why the underlying assumptions and thoughts—the problems and insights—are important for understanding the intention of each principle, meaning what problem it is trying to solve and how.

Someone posted a comment online about Agile 2, saying that if the original Agile Manifesto authors were not involved in the Agile 2 effort, he would not look at it. We believe that all great ideas build upon what has come before, and that even "original" ideas have deep roots. The term *Agile* had been used prior to the creation of the Agile Manifesto, and "agile" methods had been used and circulated for years prior to that. Not only do many Agile methods date back decades, but core ideas in Agile 2 such as Socratic leadership date back millennia.

There is also the matter of dysfunction within the Agile community, which we will discuss at length in the book. The dogma that is found in some quarters is one form of dysfunction; another is the separation of the community into tribes, for the various frameworks. We will explain why this has been a problem and how it has "frozen" Agile thinking and stifled its evolution.

Many of the thought leaders in the Agile community have a lot invested in current paradigms and practices, and so change is not in their best interest. For these reasons, we felt that we could not rely on the community to fix these problems. The problems come *from* the community—not from the whole community, but from some of the most established and entrenched parts of it.

Why bother then? Why deal with this? It's because Agile is *extremely important*. DevOps cannot replace Agile. While Agile has become mostly about the human side of building things, DevOps has become mostly a collection of technical practices. That is the reality on the ground. But there is more to building things than the technical side: one needs both the human side *and* the technical side.

We therefore realized that addressing Agile's gaps is really important, and that to do it, we needed a diverse team with a wide range of skills, composed of people who are not deeply invested in current paradigms or frameworks. We needed original thinkers. Those criteria led to the Agile 2 team of 15 members, which you can find on the Agile 2 website.

Agile 2 is an attempt to make a solid course correction to Agile, but in an open, additive, inclusive, and nondogmatic or emotional way. We welcome ideas that can supplement Agile 2 and feedback on its principles, in the spirit of inclusiveness and advancing everyone's understanding of these complex issues.

Agile 2 broadens Agile's focus beyond software. The reality is that Agile ideas have been applied for many things besides software, and so the Agile 2 team felt that it made no sense to define Agile 2 only for software.

The reader will notice that many Agile 2 principles are stated in the margin, but not all of them. This book is not a textbook about Agile 2 that covers every aspect of its principles. The purpose of this book is to introduce Agile 2, explain why we need it, and give an overview. You can find more information about Agile 2 on the website: https://agile2.net, which is published under a Creative Commons Attribution license, "CC BY 4.0."

This book attempts to make the many topics of Agile 2 concrete. We give guidance on how to apply the principles and provide examples. However, we refrain from providing specific steps or templates to follow: we do not want to repeat the mistake of current Agile frameworks in that regard.

Except for our examples, we do not define practices to implement. Practices are important, but that is for another book and for others to propose. This book lays out a conceptual foundation, while using concrete situations as examples, but not for prescription.

A note about the use of the words "agile" and "Agile": This book uses the word "Agile" when referring to the ideas embraced by the "Agile community," which is comprised of Agile coaches and others who view the agilemanifesto.org document as a guiding source of insight; or in the context of so-called "Agile frameworks" which claim to define practices that are consistent with the philosophies of the Agile community. We use the word "agile" when we intend to convey the generic quality of agility. Agile with a capital "A" and agile with a small "a" are two *different words*.

The name "Agile 2" is not intended to be a version number, as in "2.0," "2.1," etc. Rather, it is the name of a reborn Agile. We feel that the principles of agility are timeless, so we do not expect an Agile 3, and so on. Rather, we see Agile 2 as an attempt to reimagine Agile—not from scratch, but by taking Agile ideas and pivoting. We hope that Agile 2 hits closer to the mark!

1

How Did We Get Here?

At a developer conference in 2015, Dave Thomas, one of the authors of the Agile Manifesto, gave a talk titled "Agile Is Dead."[1] In a 2018 blog post, Ron Jeffries, another Agile Manifesto author, wrote, "Developers should abandon Agile."[2] In a 2019 article in *Forbes* titled "The End of Agile," tech author Kurt Cagle wrote, "[Agile] had become a religion."[3] A post about the article[4] in the programmer forum Slashdot received more than 200 comments from software developers, asserting things like "Agile does not always scale well" and "The definitions of 'agile' allow for cargo cult implementations."

Agile has been a subject of ridicule since its beginning. In the early days, there were many people who did not understand Agile and spoke from ignorance; what has changed is that today the criticism often comes from people who *do* understand Agile methods and have decided that those methods are problematic.

Is Agile actually dead? The statistics say no,[5] yet something is clearly wrong. Agile—which was sold as the solution for software development's ills—has severe problems. What are those problems, how did they happen, and what can be done about them? And is Agile worth saving?

Most of the discussion in this chapter will be about software. That is because Agile began in the software domain. In later chapters, we will broaden the discussion to product development in general, and to other kinds of human endeavor, since many Agile ideas apply to essentially any group effort.

A Culture of Extremes

In 1999 a new book called *Extreme Programming Explained* by Kent Beck sent shock waves through the IT industry. Agile ideas had been circulating and in use prior to this, but Beck's book somehow pierced corporate IT consciousness. It arguably launched the Agile movement, even though the movement was not called "Agile" yet.

The movement's core thesis was that methodical, phase-based projects were too slow and too ineffective for building software—challenging the approach then used by most large organizations and pretty much every government agency.

The book did not launch Extreme Programming, aka XP, which was first defined in 1996,[6] but it was the book that popularized it. Talk about XP could be heard in the halls of every IT shop. It was controversial, but its values strongly resonated: Small teams, working code (rather than documents) as the only real proof of progress, frequent discussions between the customer and the programmers. Out with big, up-front requirements documents that were always incomplete, inconsistent, and incomprehensible; out with big, up-front designs that were usually wrong. Recurring and incremental customer approval instead of contracts that locked everyone in to unvalidated assumptions.

Many of the methods of XP were not new, but they had been outlier methods, and XP put them under a single umbrella. The book strongly asserted that these methods work and are a superior alternative to traditional methods.

It is not that there were no other proposals for how to reshape software development. So-called lightweight methods had been around for a while. Extreme Programming was new in that it threw a grenade into much current thinking by being so radically different and proposing methods that were so *extreme*—methods such as pair programming (which had been described as early as 1953)[7] and Test-Driven Development (which also had some history prior to XP), which turned many assumptions about programming on their head.

Thus, the movement began as a rejection of the predominant existing paradigms. People knew something was wrong with software development as it was being done. Extreme Programming provided an oppositional alternative. It was not so much that people thought XP was great, but they were sure that current practices were not great. XP received a lot of attention and was a radically different approach.

Perhaps the attention was not because XP was so much better or radical, as there had been other ideas circulating such as Rapid Application Development, but perhaps XP got attention mostly because the Internet provided a new medium that made rapid awareness possible.

Then in 2001 a group of IT professionals—all men by the way, with most from the United States and a few from Europe—got together over a weekend and hammered out a set of four "values," which they believed should be the foundation of a new approach to building software. Kent Beck was among them. You can find these four values at AgileManifesto.org. It was largely a rejection of many approaches that had become commonplace, such as detailed plans, passing information by documents, and big all-at-once deliveries.

In the weeks that followed, some of them continued the discussion by email and added 12 principles, which you can also find at the same website.

They called all this the Manifesto for Agile Software Development, and it came to be known colloquially as the Agile Manifesto or just Agile. This "manifesto" took the popular culture baton from XP and other iterative approaches and launched the Agile movement for real.

Extreme Programming had set the tone for what would become the Agile movement, and the tone was to be extreme. In those days, *extreme* was popular. We had extreme rock climbing, extreme skateboarding, extreme pogo, extreme skiing, and extreme pretty much anything. Extreme was in. People were so tired of the ordinary; everything new had to be extreme. It was a new millennium for crying out loud: everything needed a reset!

And so "extreme" was a necessary aspect of anything new and interesting at that moment in time in the late 1990s—the end of the 20th century.

Since Agile was a rejection of what had become established software development methods, it was inherently a disruptive movement, and in the ethos of the time, it had to be extreme. And so it was that every Agile method that came to be proposed—these are called *practices*—were of necessity extreme. Otherwise, they were not seen to be consistent with the spirit of being entirely new and disruptive.

It was not the Agile Manifesto that set things in that direction. The Agile Manifesto was clearly about balance and moderation. It makes no absolute statements: every value is couched as a trade-off. For example, the first value reads, "[We have come to value] individuals and interactions over processes and tools."

It does not say, "Forget process and tools—only pay attention to individuals and interactions." Instead, it says, consider both, but pay special attention to individuals and interactions.

In other words, the Agile Manifesto advocated judgment and consideration of context. In that sense, it is a sophisticated document and cannot be used well by people who do not have the experience needed to apply judgment.

But the tone had already been set by XP: extreme practices received the most attention and applause, because XP practices were all extreme. For example, XP's recommendation of pair programming, in which two people sit together and write code together, sharing a keyboard, was considered by many programmers to be extreme. Or everyone sitting side by side in a single room, with all walls removed and no privacy—that was pretty extreme, as it had been assumed that people needed privacy to focus, and the big programmer complaint of the 1990s, depicted in so many Dilbert cartoons, was that programmers were no longer being given offices and instead were being sat in cubicles that did not afford enough quiet or privacy. And now here comes XP and says, in effect, *You got it all wrong; you need to sit next to each other*. That was an extreme swing of the pendulum.

The Scrum framework, which dates in various forms to the 1980s but became reformulated in 1993 and then popularized through its certification regime during the 2000s,[8] added more ideas that are arguably extreme. For example, Scrum views everyone on a team as an equal player—no one is acknowledged as having more standing than anyone else ("Scrum recognizes no titles for Development Team members"[9]), regardless of their experience. That was pretty extreme, since before Agile, programmers in most (not all) organizations had professional levels, such as programmer, senior programmer, architect, etc.

During the first decade of the Agile movement it seemed that new suggested practices were in a competition to be more extreme than the others. We saw the introduction of mob programming, in which rather than two programmers working together as in pair programming, the entire team works together—literally—everyone calling out their thoughts in a single room and sharing a single keyboard.[10] Then in 2015 the Agile team room was extended by Facebook to an extreme level when it created its 430,000-square-foot open team room.[11]

We also saw the growth of conference formats pushing popular Agile practices to the extreme—for example, the Lean Coffee format in which there is no agenda or the "unconference" and Open Space formats in which people vote on topics and join ad hoc conversations. These contrast strongly with a traditional conference or discussion group format, which generally has an agenda and scheduled talks, with informal sessions afterward.

But do extremes work?

Certainly they work for *something*. There is always some use for any tool. The question is, are the extreme practices advocated by many among the Agile community actually the most effective method for a wide range of situations that commonly occur in organizations? Another question is, do these practices favor certain ways of working at the expense of others so that certain people benefit but others are at a disadvantage? Or, should a thoughtful approach be used, with moderate approaches being the norm and extreme approaches used sparingly and when a particular form of activity is desired for a specific reason?

Since the Agile movement began through advocacy of extremes and was inherently a disruptive movement, it became evangelistic, and dogmatic elements arose. As a result, if one did not embrace the trending favored set of Agile practices, one was at risk of being labeled an "Agile doubter." That label was brandished readily by many Agile coaches. And so extremes came to be not just something to consider, but *the* way—and the only way. Agile was now something to accept on faith; as Kurt Cagle had written, it had become a religion.

Divided and Branded

Whenever something new and useful is created, people and organizations jump in to claim it and use it for their own purposes. Any change creates huge opportunities. For example, when Howard Head came out with the oversized Prince tennis racquet in the early 1970s, other manufacturers followed Head's lead and came out with racquets that departed from the standard size. One suddenly saw racquets on the market with very large nets and also ones that were only slightly larger than what was then the standard size.

The ones that were slightly larger became the most popular and came to be seen as the new standard size. These were called "midsize" racquets, although today we just call them racquets.[12]

The inevitable result was that everyone who owned a tennis racquet had to go out and buy a new one; otherwise, they were not adhering to the new "standard." The sports equipment industry experienced a windfall in sales.

The rise of Agile did the same thing. There were new books, new websites, new consulting practices, and new frameworks that purported to be Agile. Agile quickly became a commodity to sell. It began with an Agile certification industry. Ken Schwaber introduced a two-day certification course for his Scrum framework: the Certified Scrum Master, aka CSM. By sitting in a training room for two days and not even taking a test (they do provide a simple test now), one could walk away with a certificate claiming to be a "master."

Essentially the material that could be contained in a small pamphlet was the basis of what Human Resources staff and many hiring managers erroneously interpreted as a "master-level" certification.

Since Schwaber and his partner, Jeff Sutherland, claimed that Scrum was an Agile framework, it was something they could sell under the rising banner of Agile. Organizations that preferred to hire people with certifications made the CSM a requirement. People who had master's degrees from universities were dismayed at the naively perceived equivalence of a master's degree and a Certified Scrum Master certification. Highly qualified people were screened from job applications because they did not have the two-day CSM certification.

Industry groups sprang up: the Agile Alliance, the Scrum Alliance, Scrum.org, ICAgile, SAFe, LeSS, Kanban, and many others. The large consulting companies had a hard time learning about Agile, because Agile's central message of being lean and efficient and not having big contracted "phases" was antithetical to their model. Eventually they figured out how to incorporate it into their offerings, and today they all have substantial Agile practices, claiming to be the experts in Agile.

Thus, there is a lot of money today in the Agile industry, and the Agile community is arguably driven by moneyed interests, with a continuing tide of people seeking certifications in the various frameworks and becoming indoctrinated into them. The phrase *Agile industrial complex*, which might have been coined by Martin Fowler (one of the authors of the Agile Manifesto), has come to be used to refer to the industry as a commentary on the degree to which it is driven by financial interests rather than by ideas and efficacy.

Controlled by Dogma

Today Agile is big business, and it is highly competitive.

Large consulting companies have traditionally used their partners to fly around and build relationships with their clients' executives, convincing the executives that the partners and consulting firms have strategic insight. That was the case before COVID-19 and will probably resume being the case after COVID-19. Through those relationships, the partners are able to place large numbers of Agile-certified staff on-site, generating a lot of revenue.

Placing staff is their goal—as many as possible. These staff members do not usually have the industry experience that the partners have, but they have a certificate, perhaps a Certified Scrum Master certificate, perhaps a SAFe certificate, or perhaps others. In other words, they sat in a classroom for a few days or weeks, and they probably participated in at least one project that was said to be Agile. Most are not what one would normally expect from a consultant who is advising teams and programs: decades of practical experience at multiple levels of responsibility, great acumen, demonstrated industry thought leadership, and a history of P&L accountability.

Certainly there are many people in large consulting companies who are very qualified, but we have seen many who are not. The point is that one should not treat a large consulting company as if it is inherently more trustworthy than any other with regard to Agile expertise. Have their people achieved business results? The partners will tell you they have and will provide proof, but you know how data can be misused.

Many of the various industry groups do not like each other either. One of the largest Scrum training organizations—one that claims to speak for the Scrum community at large—actually writes into its contract with its trainers that one cannot train for a "competing" training framework such as the Scaled Agile Framework. Thus, the organization does not want its customers to have a choice or to be able to consider other ideas. Limiting what one's trainers know does not seem to us like a good way to serve one's customers.

Meanwhile, the organization preaches Agile's value of "customer collaboration over contract negotiation" and professes to being open to all ideas and to being cooperative and collaborative—core Agile attributes. You should not miss the hypocrisy in this.

An Agile training organization's armies are its certified coaches. When the CSM certification was introduced, it had an unintended effect (at least, there is no evidence that it was intentional). Large numbers of people who had no software development experience took the training and became certified. Almost overnight they had a new career: their CSM certification could get them a job as a Scrum Master or an Agile coach.

We then saw the rise of these two new professions (Scrum Master and Agile coach), and they quickly came to be dominated by nontechnical people from every manner of prior profession. Since their background was nontechnical, they emphasized the human side of software teams: team trust, team happiness, and so on, as well as the set of Scrum practices: the sprint planning, the daily scrum (aka daily standup), the sprint review, and the retrospective. The technical focus that XP had was almost entirely lost, and by 2011, 52% of Agile teams reported using Scrum (which defines no technical practices), compared to only 2% using the highly technical XP.[13]

This is a big problem. In a talk at Agile Australia in 2018, Martin Fowler, one of the authors of the Agile Manifesto, asked the audience, "How many people here are software developers?" Then he said, "A smattering, but actually very much a minority . . . and that's actually quite tragic."[14]

Agile coaches and Scrum Masters focused on the nontechnical practices as if they were the be-all and end-all of software development, ignoring critical things such as test coverage, code branching, integration testing, issue management, and all the critical things that every programmer knows are essential. The retrospective, in which a team is supposed to talk about how to improve their work, was facilitated by the Scrum Master, who was usually nontechnical, and so the discussion tended to steer toward the Scrum practices, because team members did not want to bring up issues that the Scrum Master would not understand. Teams then went back to their desks, anxious to get back to work after all of the Scrum-related ceremonies (what the Scrum practices are called), and so important technical issues went undiscussed.

The Agile conversation was essentially taken away from software developers—the people for whom Agile was created. After all, the Manifesto begins with (italics added), "*We* are uncovering better ways of developing software . . . ."

Agile thought leadership was increasingly controlled by those who wanted to advance an increasingly extreme behavioral agenda. Programmers speak up from time to time, but with trepidation. For example, a poster on Reddit wrote this:

> *"I hate Scrum. There. I said it. Who else is joining me? Scum seems to take away all the joy of being an engineer. working on tasks decided by someone else, under a cadence that never stops. counting story points and 'velocity'. 'control' and priority set by the business - chop/change tasks. lack of career growth—snr/jnr engineers working on similar tasks."[15]*

The Scrum Master and Agile coach roles became so entrenched that organizations now assume that they need those roles. No one questions it. More important, no one questions the skills and knowledge that are needed for those roles to actually be effective.

The career of a Scrum Master or Agile coach relies on the perceived value that they provide. Since the experience of most was not in the technical realm, the natural evolution was to inflate the value of the nontechnical dimension of things. We saw a rise in Agile dogma, in which Agile or Scrum advocates—usually Agile coaches or Scrum Masters—would deride anyone who challenged Scrum or Agile practices. We have mentioned the term *Agile doubter*. These dogmatic people were such a problem that in a 10-year retrospective on the state of Agile, the British Computer Society referred to Scrumdamentalism—a play on the words *Scrum* and *fundamentalism*—as one of the main problems with the state of Agile.[16]

The Agile community suffered from groupthink: the community mostly spoke to itself. One of our editors described the community as insular. During the first decade of Agile, Agilists mostly read books that said "Agile" on the cover, but important ideas from other communities, such as those that study organization change or organizational psychology, did not permeate the Agile community. In the words of *Norwegian Wood*'s author Haruki Murakami, "If you only read the books that everyone else is reading, you can only think what everyone else is thinking."

The result was that thinking in the Agile community became highly constrained: the community would only tolerate ideas that either modified existing Agile practices to be more extreme or added new nontechnical practices that were also extreme. Anything else was either ignored or dismissed as "not really Agile." So for the entire decade of the 2000s, Agile was confined largely to nontechnical thinking—an irony since Agile was created for software development.

Meanwhile, companies such as Google, Amazon, and Netflix had real problems to solve. They needed to be able to deploy to hundreds of millions of users many times a day, with confidence, and allow teams to make rapid changes and redeploy without delay. They needed agility in their software development and release processes.

So, they invented techniques to help them to achieve these things: techniques such as on-demand test environment creation, test-first integration testing (aka ATDD), containerization, container cluster management, and many other things. Jez Humble was the first who we know to have assembled these ideas into a coherent whole and publish a book about them, which he called *Continuous Delivery*, and the methods he described became known as *continuous delivery*, often abbreviated as CD. Gene Kim later wrote about CD ideas in his book *The Phoenix Project*.

Humble was an advocate of Agile methods and viewed himself as an Agilist, and he spoke about his work at Agile conferences. His talks were well attended, but few in the Agile community at large paid attention: what Humble was advocating was technical, so it was largely ignored by the large and growing cohort of nontechnical Agile coaches, particularly those who had built a career around Scrum. The established spokespeople for Agile—the popular Agile authors of the day—also largely ignored what Humble was talking about, preferring to pile onto the process topics that had been trending, and a few kept tweaking the now dated practices of XP. As a result, CD methods and related ideas became their own movement, which we know today as DevOps—a term that arose shortly before Humble's book was published.

The Agile community saw the rise of DevOps in the early 2010s and became alarmed. Agile was under threat from something new, so they claimed it as their own. Those who still advocated for XP claimed that DevOps was just XP in a more evolved form (not true). Agilists pointed to Humble's Agile origins but did not acknowledge that the majority of Agile coaches paid little attention to CD until DevOps rose to widespread awareness.

To be fair, there are many technical Agile coaches, even though they are a small minority. Those people indeed saw CD as continuation of the march of Agile ideas that began with XP and before. They were a fractional cohort, though.

Today if you ask an Agilist about DevOps, they usually think that DevOps is an evolution or outgrowth of Agile. That is not how it was, but it no longer matters. Agile is a set of ideas. DevOps is a set of ideas. Both sets of ideas, and many others, are extremely valuable in the context of product development.

The Introvert vs. Extrovert Problem

Earlier we posed the question, do these practices favor certain ways of working at the expense of others so that certain people benefit but others are at a disadvantage?

The group that authored the four values of the Agile Manifesto was pretty homogeneous. As we said, they were all English speaking, most were from the United States with a few from England and one from the Netherlands, all were men, and all were very experienced. They did not represent a "typical" team of programmers. It is reasonable to expect, therefore, that they had some collective biases that differ from how a typical team of programmers would feel and think.

Since most were experienced, one might assume that when they wrote that they value "individuals and interactions over processes and tools," they were thinking of themselves as capable individuals—individuals who are highly experienced and who have refined judgment about IT methods. Such individuals arguably need less oversight—less *process*.

Within the principles that some of the group came up with, one of them reads, "The best architectures, requirements, and designs emerge from self-organizing teams."

That one principle has resulted in a strong preference within the Agile community for "self-organization," that is, a team in which there is no designated leader. The authors of the Agile Manifesto were able to come up with four values in the course of a weekend, so they clearly were able to self-organize enough to accomplish that. Would they have been able to continue to be organized over a period of months to create a complex software product?

No one can say. The principle regarding self-organizing teams has been one of the most contentious statements of the Agile Manifesto. An article by one of us in LinkedIn about self-organizing teams received 45,000 reads in the first two weeks—two orders of magnitude higher than the normal read rate for that author.

What happens if you put a group of people together and tell them to "self-organize" to achieve a goal? Well, it depends.

You will read that phrase, "it depends," a lot in this book. We believe it is central to all of the issues pertaining to how groups of people should work together. In the words of Malcolm Gladwell, the author of *Blink*, the only answer that is always right is "it depends," and that is definitely true when dealing with groups of people.

Some of the factors affecting how well a group self-organizes are:

- **Personalities**—some personalities mix well, others less so.
- **Level of knowledge and experience**—do they know the job well enough to get it done without supervision?
- **Urgency**—how important is their goal? Do their lives depend on it? Or at the other extreme, is it something that someone else wants—something the team sees as inconsequential to them— and they are mere mercenaries?
- **Preparation**—have they been trained to work in a certain way so that everyone has predefined roles?
- **Proven history**—have they worked together before and shown that they can?

There are certainly other factors as well.

Some people work well in a group; others have trouble in a group and prefer to have their own task. The Agile community tends to dismiss the value of these "loners" or "lone wolves." Yet lone wolves have created some of the most impactful software. Linux is an example. It was created by Linus Torvalds, who is a self-proclaimed loner and introvert. Today he still oversees the evolution of Linux, but he does so in a room by himself, communicating with the thousands of Linux kernel developers involved entirely by email. Linux powers most of today's computers, from smartphones to most of the servers in the cloud. If you use Amazon, you use Linux. If you use an Android phone, you use Linux.

Introverts tend to shy away from groups. Groups are taxing for them. In contrast, extroverts seek out groups—groups energize them.

It follows then that in a team in which the prescribed form of communication is face-to-face and verbal, introverts would feel a lot more taxed than extroverts, who would actually find the frequent face-to-face discussions energizing.

One of the principles in the Agile Manifesto reads, "The most efficient and effective method of conveying information to and within a development team is face-to-face conversation."

This seems to clearly favor extroverts. Extroverts will enjoy resolving every issue by turning around and talking to others. They would enjoy convening an ad hoc meeting to discuss the issue. According to Noa Herz, a neuroscientist and neuropsychologist, "Group meetings, in which each participant contributes thoughts in a disorganised, dominance-based manner, can put introverts at a disadvantage."[17]

An Agile coach or Scrum Master is supposed to facilitate team meetings, but doing so can be challenging. If a few members are talking rapidly, a facilitator might feel that the members are "on a roll" and be hesitant to interfere.

An introvert—by definition—will find that situation emotionally taxing. Even if asked their opinion, they are not likely to keep the floor long enough to actually convince anyone. It is therefore reasonable to assume that introverts might prefer to first communicate in written form and meet in person only if it is necessary.

An interesting fact about programmers as a group is that they *think* they are extroverts.[18] Yet some psychological studies indicate that programmers actually tend to be introverts.[19] It is a complex issue, though, because other studies produce conflicting results on the question.

Regardless, there is still a problem, because while Agile appears to favor extroverts, the people at whom Agile methods are mostly targeted—software developers—have a high proportion of introverts, as a group, even though the actual percentage is unclear. Even if, say, 60% of programmers are extroverts—something that would be a counterintuitive career choice—forcing extrovert-favoring methods on the other 40% would be unfair and counterproductive.

It is also a problem because by sidelining introverts, teams suffer. According to Herz, "Entrepreneurial teams perform better when leadership is shared between individuals, but only if they have diverse personality traits. Moreover, teams dominated by extraverted members actually perform better under introverted leaders,[20] possibly due to their greater responsiveness to their employees' ideas."

How did the Agile Manifesto come to be written to favor extroverts? Were the authors of the Agile Manifesto extroverts? It is hard to say, but the manifesto's stated preference for face-to-face communication aligns with what one of the authors had been writing shortly before their meeting. Alistair Cockburn had written extensively about how, in his opinion, face-to-face communication is the "most effective" form of communication.[21]

So it is possible that one member of the group influenced the others—a member who seems to fit the profile of an extrovert, based on his high level of involvement in Agile community group initiatives.

Is it really true? Is face-to-face communication always best?

It depends. (There it is again.) We will dig into that question a great deal in Chapter 6, when we talk about how collaboration occurs for simple and complex issues.

What about the people who help Agile teams to define how they work? That is, Agile coaches and Scrum Masters? Do they tend to be introverts or extroverts?

We are not aware of a study on that question, but consider that in seeking such a job, one is looking for a role in which one is coordinating groups of people and encouraging face-to-face communication. It stands to reason that the job, as defined by Scrum and the Agile community, would attract people who like working with groups of people—in other words, extroverts. And it has been our experience that Agile coaches do, by and large, seem to be "people people," although they are as varied as the people in any profession, and no two are the same.

If it is true—if Agile coaches and Scrum Masters lean slightly extroverted—then we have a slightly extrovert-leaning population advising a somewhat introvert-leaning population on how to work. What could go wrong?

Ignoring whether an Agile coach or Scrum Master is an introvert or extrovert or something in between, and following default Agile practices that favor extroversion can be detrimental if those practices are not what are actually preferred by the team members.

Does this matter? Is there actually a problem? What impact might this have?

Let's look at common complaints that we have heard from programmers who work on Agile teams.

- "I can't focus! There are too many distractions and disruptions."
- "Too many meetings!"
- "I don't actually *want* to know what everyone else is working on!"
- "I don't actually *want* to self-organize! I want someone to *get us organized* so that I can code! I just want them to let me *code the way that I want to*."
- "I don't actually *want* to reach out to others for ad hoc discussions! I only do that as a last resort if I am really stuck."
- "Not enough attention is paid to technical issues!"

If these complaints are valid, it is a pretty bad state of affairs, because the true needs of programming teams are not being addressed by the Agile community; in fact, it is worse than that: programmers are being literally coerced, by organizations that adopt Scrum as a mandatory methodology,[22] into working in a way that is not suited to their personalities.

Coaches Should Not Assume

What has continued to amaze us is that if one asks an Agile coach, "Is Agile working?" the answer will usually be an emphatic yes followed by gushing about how productive teams are and how Agile has revolutionized software development. Yet if one asks the typical programmer—the very people who Agile was created for—they will have a different range of responses. In fact, it seems to us that most programmers have not had a very positive experience with Agile.

Studies at the University of Florida and the University of Notre Dame indicate that introverted team members tend to view their extroverted co-workers more negatively, whereas the extroverted co-workers do not seem to have that negative bias.[23] In other words, the introverts silently judged the extroverts negatively.

We think something like that might be happening among Agile teams. Agile coaches tend to be people-oriented, and therefore more extroverted than programmers on average, so they don't see a problem; programmers tend to be more introverted than average.[24] The Agile coaches think things are great. The programmers? Well, not so much.

The University of Florida and Notre Dame experiments might also indicate that extroverts tend to be relatively insensitive to how introverts are experiencing things, because if they were aware, they would probably not be so favorable in their own ratings of their introverted colleagues. That conclusion is also supported by a Yale study that indicated the introverts are generally more sensitive to the feelings of others.[25]

This is why coaches should not assume that what *they* would prefer is what their team will prefer. Ask! And pay attention to what is being written and said, even if it goes against Agile dogma. For example, in her book *Carved In Sand*, Cathryn Ramin writes the following on page 33:

> "... an endless stream of interruptions has become the norm. Researchers at the University of California, Irvine attempted to quantify the number of distractions and interruptions that occur among IT workers in a medium-sized office. They predicted that something would interfere with concentration every fifteen minutes, but on average, interruptions occurred every three minutes, and only two-thirds of the interrupted work was resumed on the same day."

If this sounds like your organization, do you think that team members can focus? Ask them!

What about use of tools like Slack and Microsoft Teams? Are they too distracting? Consider this article in Nir & Far, "If Tech Is So Distracting, How Do Slack Employees Stay So Focused?":

> "Slack management leads by example to encourage employees to take time to disconnect. In an interview with OpenView Labs,[26] Bill Macaitis, who served as Slack's chief revenue officer and chief marketing officer, states, "You need to have uninterrupted work time . . . . This is why—whether I'm dealing with Slack or email—I always block off time to go in and check messages and then return to uninterrupted work."[27]

Perhaps do not assume that the team should be texting each other all day in these tools. Perhaps encourage them to disable desktop notifications and to not install the tool on their phones.

Maybe some people are bothered by these tools and by ambient noise, but others are not. Ask! Consider this post in Slashdot, "Why Office Noise Bothers Some People More Than Others":

> *"According to a 2015 survey of the most annoying office noises by Avanta Serviced Office Group, conversations were rated the most vexing, closely followed by coughing, sneezing and sniffing, loud phone voices, ringing phones and whistling. Why do we find it so hard to be around these everyday noises? . . . As the researchers suspected, all the students performed better in silence. But they also found that . . . the more extroverted they were, the less they were affected by noise."[28]*

Maybe some team members can work just fine in a team room, but others cannot. Do not assume that a team vote is the answer. Maybe a range of approaches should be considered. Ask!

Noise and pop-up alerts are not the only problem. Some people also find local movement distracting. That is why Panasonic has developed "horse blinders" for humans, specifically for use in open office layouts (Figure 1.1).[29]

Figure 1.1: Blinders to help people focus in "team rooms"

Source: (Photo from Japan Trends:
www.japantrends.com/wear-space-panasonic-wearable-concentration/)

Are frequent "stopping by" discussions disruptive? For some people, yes. According to a study at George Mason University, being interrupted for only 60 seconds while concentrating can completely wipe one's short-term memory.[30]

Perhaps do not advise your team to just "stop by" each other's desk at any time. Perhaps people should have an "I'm thinking" sign on their desk, indicating that they do not want to be disturbed.

What about all the Agile ceremonies? Many developers think it is too much.[31] Perhaps instead of assuming that the team needs those, ask! According to an article in Doist.com,

> *"This trend toward near-constant communication means that the average knowledge worker must organize their workday around multiple meetings, with the time in between spent doing their work half-distractedly with one eye on email and Slack."*[32]

Talk to teams. Find out how they feel that they work best. Do not assume. Do not blindly follow someone else's methods.

Now What?

We can see that while there are some really powerful and important ideas behind the Agile movement, things are not quite right. What should we do?

Agile 2 is not intended to be a new set of ideas. The ideas of Agile 2 are all well-established, but they are not Agile doctrine. Many members of the Agile community have been using these ideas on an individual basis but usually describe these as implied nuances, or "bringing in ideas from other domains." These ideas pertain to leadership, to effective collaboration, to empowering people to work in a way that is best for them individually, and to applying context and judgment instead of just adopting an extreme practice. Agile 2 is about bringing these nuances explicitly into the Agile idea set and making them part of the Agile envelope.

In the next chapter, we will take a more comprehensive look at some problems of Agile to establish a better understanding of what areas of Agile need improvement. In later chapters, we will then look at models for effective leadership, collaboration, product design, transformation, continuous delivery, and other ideas that are part of Agile 2, and we will conclude with examples of some specific problem domains through an Agile 2 lens.

Notes

1. www.youtube.com/watch?v=a-BOSpxYJ9M
2. ronjeffries.com/articles/018-01ff/abandon-1/
3. www.forbes.com/sites/cognitiveworld/2019/08/23/the-end-of-agile/
4. developers.slashdot.org/story/19/08/24/1748246/is-agile-becoming-less-and-less-relevant
5. goremotely.net/blog/agile-adoption/
6. www.extremeprogramming.org/
7. www.oreilly.com/library/view/making-software/9780596808310/ch17s01.html
8. en.wikipedia.org/wiki/Scrum_%28software_development%29#History
9. www.scrumguides.org/scrum-guide.html#team-dev
10. www.agilealliance.org/glossary/mob-programming/
11. www.fastcompany.com/3044389/a-first-look-at-facebooks-new-mothership-designed-by-frank-gehry
12. The larger racquets were not welcomed by many tennis enthusiasts, because the larger racquets encouraged sloppy strokes, and they also produced a faster serve, shifting the focus of the game to one's serve.
13. explore.digital.ai/state-of-agile/6th-annual-state-of-agile-report
14. www.youtube.com/watch?v=G_y2pNj0zZg&feature=youtu.be
15. www.reddit.com/r/devops/comments/i4cbwu/i_hate_scrum/
16. www.slideshare.net/jcasal1/20120419-agile-business conferencepptx, slide 11.
17. psyche.co/ideas/introverts-are-excluded-unfairly-in-an-extraverts-world
18. www.infoq.com/news/2013/02/Introverted-Intuitive-Logical/
19. ir.lib.uwo.ca/cgi/viewcontent.cgi?article=1005& context=electricalpub
20. psycnet.apa.org/record/2011-15936-006
21. www.researchgate.net/figure/14-Effectiveness-of-different-modes-of-communication_fig8_27296733

22. The Scrum authors refer to it as a *framework*, but Scrum perfectly fits the definition of *methodology* as a "body of methods, rules, and postulates employed by a discipline" (Merriam-Webster). A methodology can be extended, just as a framework can.

23. www.eurekalert.org/pub_releases/2014-12/osu-ics121614.php

24. ir.lib.uwo.ca/cgi/viewcontent.cgi?article=1005&context=electricalpub

25. www.inc.com/glenn-leibowitz/yale-psychologists-introverts-are-better-than-extroverts-at-performing-this-essential-leadership-skill.html

26. expand.openviewpartners.com/former-slack-cmo-bill-macaitis-on-how-slack-uses-slack-868ffb495b71

27. www.nirandfar.com/slack-use/

28. slashdot.org/story/363608

29. techcrunch.com/2018/10/17/open-offices-have-driven-panasonic-to-make-horse-blinders-for-humans/

30. www.dailymail.co.uk/sciencetech/article-2697466/Do-not-disturb-Being-interrupted-just-60-seconds-concentrating-completely-wipe-short-term-memory.html

31. medium.com/serious-scrum/time-spent-in-scrum-meetings-75e38b08d8

32. blog.doist.com/asynchronous-communication/

2

Specific Problems

The first thing that the Agile 2 team did was discuss the state of Agile today; in other words, we conducted a kind of retrospective. We are a global team of 15, spanning time zones from California to Vietnam, so we could not assemble in person or even remotely at the same time. We had to find another way. We had one-on-one discussions, which were documented and shared in written form, and then we collaborated to develop a set of "key ideas," which were ideas expressed by at least two Agile 2 team members. Many of these key ideas were problems with the state of Agile. Others were insights: realizations about why something did not work or what is needed to make things work better.

Let's look at some of the problems that we found with Agile today and at some insights about them.

Leadership Is Complex, Nuanced, Multifaceted, and Necessary

In the Agile community, the term *leadership* in the context of a team is often thought of in terms of the Scrum Master role, which is defined by the Scrum framework. There is also the Scrum Product Owner, who provides leadership with regard to the product vision and its feature set; and Scrum team members are expected to apply individual leadership when collaborating and organizing their collective work. Let's consider the Scrum Master role, which is the primary leadership role defined by Scrum pertaining to the work process of a development team.

The Scrum Guide's definition of the Scrum Master role has changed greatly over the years. In early 2007 the Scrum Guide said[1] that *the Scrum Master "is responsible for the Scrum process, for teaching it to everyone involved in the project, for implementing it so it fits within an organization's culture and still delivers the expected benefits, and for ensuring that everyone follows its rules and practices."*

In other words, back then, the authors of the Scrum Guide, Ken Schwaber and Jeff Sutherland, viewed the Scrum Master as mainly a master of ceremonies. The part about "implementing it so it fits within an organization's culture" is pretty ambiguous, but it sounds like it pertains to tailoring Scrum in some way.

Then in March 2007 they added three responsibilities to the role, which we paraphrase here:

- The ScrumMaster needs to know what tasks have been completed, what tasks have started, and any new tasks that have been discovered, and so on.
- The ScrumMaster needs to surface dependencies and blockers that are impediments to the Scrum team. These need to be prioritized and tracked.
- The ScrumMaster may notice personal problems or conflicts within the Scrum that need resolution. These need to be clarified by the ScrumMaster and be resolved by dialogue within the team, or the ScrumMaster may need help from management.

This adds some pretty specific process responsibilities, akin to task management, which is strange, because Agilists usually prefer stories over tasks. The reason is that a *story* is generally considered to be something that is defined in terms of its desired outcome, whereas a *task* is a prescriptive statement of what to do and therefore how to achieve the outcome. It is not that Agile teams never work on tasks, but if something can be expressed in terms of a desired outcome, that is preferred.

Then in 2009 they added (emphasis added here) the following:

The ScrumMaster is a **facilitative team leader**. . . . *He must: Ensure that the team is fully functional and productive; Enable close cooperation across all roles and functions; Remove barriers; Shield the team from external interferences; and Ensure that the process is followed. . .*

Then in 2010 they changed the role to be as follows:

> *The Scrum Master is* **responsible for ensuring that the Scrum Team adheres to Scrum values, practices, and rules**. *The Scrum Master helps the Scrum Team and the organization adopt Scrum. The Scrum Master teaches the Scrum Team by coaching and by leading it to be more productive and produce higher quality products. The Scrum Master helps the Scrum Team understand and use self-organization and cross-functionality. The Scrum Master also helps the Scrum Team do its best in an organizational environment that may not yet be optimized for complex product development. When the Scrum Master helps make these changes, this is called "removing impediments."* **The Scrum Master's role is one of a servant-leader** *for the Scrum Team.*

Note in particular the use of the term *servant leader.*

That's a lot of churn for a key role (Scrum defines only three roles), and it has changed again. As of this writing, the Scrum Guide says that the Scrum Master role is responsible for the following:

- Coaching the Development Team in self-organization and cross-functionality
- Helping the Development Team to create high-value products
- Removing impediments to the Development Team's progress
- Facilitating Scrum events as requested or needed
- Coaching the Development Team in organizational environments in which Scrum is not yet fully adopted and understood

It says other things as well, but this is the part we want to focus on, because it pertains to leadership. It basically says that a Scrum Master is a kind of facilitator, impediment remover, and advocate for Scrum. However, the second bullet is extremely ambiguous: "Helping the Development Team to create high-value products." That could mean anything! It could mean that the Scrum Master rolls up their sleeves and writes code. So that one tells us nothing, and we shall ignore it.

One thing to note about all of these definitions of the Scrum Master role is that it is focused on a single team. Team leadership is one of many levels of leadership in an organization, and so the Scrum Master role cannot be assumed to be a model for leadership at other levels of an organization.

Another thing to note is that it is very inward-focused: it is about helping the team. The outward-focused elements are the ones that pertain to advocacy for Scrum, such as "Coaching the Development Team in organizational environments in which Scrum is not yet fully adopted and understood."

A major function of the Scrum Master role is to remove impediments for their team. What does it take to do that? Generally, it means that the Scrum Master must meet with other stakeholders in the organization and persuade them to make changes to how they operate. Anyone who has ever worked in a large organization knows that getting people's attention usually requires that you have some kind of "standing." Your standing can come from being responsible for something, from having authority over something, or from a reputation of respect that precedes you. If you don't have standing, people don't pay much attention to what you want; they have urgent priorities and too little time.

A Scrum Master's standing is that they represent a team, and the team is responsible for delivering something. But a Scrum Master is severely handicapped because they cannot make an agreement on the team's behalf. Negotiating with someone usually requires that if they agree to do something, then you agree to do something that they want in return. A Scrum Master cannot do that: their only tool of persuasion is that their team needs it, and that is all.

In other words, a Scrum Master lacks any kind of authority, except that they can prescribe that the team must adhere to Scrum. Their lack of authority makes them unable to make promises or deals on their team's behalf.

We also saw that the Scrum Master role is entirely focused on a team—a single team. In a large organization, there are many other contexts in which leadership is needed. As Peter Drucker said in his book *The Practice of Management*, an organization needs an "outside person," an "inside person," and a "person of action." Drucker was describing different styles of leadership.

Leadership is a complex and nuanced issue. We find that the Agile community over-emphasizes self-organization and, in doing so, over-simplifies leadership, often dismissing the need for any kind of explicit leadership, as if good leadership always emerges from a team on its own. In fact, we feel that leadership is the central issue that needs to be understood well and not over-simplified. With good leadership, everything else follows.

We also do not believe that there is a way to perform "leadership by numbers"; that is, there is no organization design or process that can sidestep the need for good leadership. People have tried to do that: to prescribe processes that avoid the need for authority. However, what happens is that individuals attain influence, and thereby de facto authority, and one is back to having individual leaders again.

The problem of needing good leadership will not go away. It must be met head-on. The Agile movement has dismissed explicit leadership: the manifesto's stated preference for a self-organizing team was taken to mean that explicit authority is not needed. Yet to form a team in the first place, one needs authority. We will delve more into this topic in Chapters 3 through 5.

The Scale Problem

From the beginning, Agile ideas were expressed in terms of "the team," implying that there is one team working on one product. A one-team product occurs frequently, but multiteam products are just as common, if not more so.

Could it be that Agile methods only work for a single team? Maybe Agile methods worked well, not because the methods were good but because they were commonly applied for the easy case: one team. We do not believe that, but it is a valid challenge.

There is also the case of multiple teams but a single monolithic product. In the early days of Agile, most software products were monolithic; that is, they consisted of a small number of very large bodies of code. A typical architecture was a web application, which consisted of a user interface web page, a web server, and an application server.

That kind of architecture could become pretty complex, and there were frameworks for breaking up the application into pieces on the server. The most well-known examples were the Enterprise JavaBean architecture and the web service architecture, and these patterns were often used together. Still, the number of runtime components was generally a handful at most.

Contrast that with today. A typical architecture at, say, a large bank, is that the company's digital platform will consist of tens or hundreds of products, each supported by hundreds of distinct microservices, and all these microservices interact in various ways in real time by sending messages or calling each other. There are often hundreds of development teams working on these many moving parts, which together constitute an integrated product ecosystem.

In his *Forbes* article "The End of Agile," Kurt Cagle wrote this:

> *Scale problems only show up once you've built the system out almost completely and attempt to make it work under more extreme conditions. . .This becomes even more of an issue when developers have to integrate their efforts with other developers, especially for those components developed at the same time.*[2]

The complexities of scale are what make the scale issue perhaps the most important technical issue today. It is also an issue that strongly overlaps with the issue of how leadership should scale: through a traditional hierarchy, through a network, through informal means, or in other ways.

Single-team startups generally had little trouble using Agile methods. Frankly, a startup is the easy case; it is "table stakes" for Agile. The hard case is making Agile work for a multiteam product, and an even harder problem is making it work in a multiproduct ecosystem.

Agile had no answers for these situations in the early days. Over time, various people tried to address the issue, by defining Agile scaling frameworks, but unlike the original Agile Manifesto, these attempts generally came from individuals and were used as brands to drive proprietary consultancies, so no consensus emerged on what to do.

Meanwhile, large consultancies embraced one or more of the branded scaling frameworks, having their staff obtain certifications in those and then marketing their staff as commodity experts (which seems to us like an oxymoron). This further entrenched the commercial nature of Agile and the approaches for "scaling" Agile. To say that the situation became unhealthy is an understatement.

Today's Tech Platform Is As Strategic As the Business Model

The Industrial Revolution brought about large companies that built really big things—things that required machines. Carnegie Steel, Rockefeller's Standard Oil, Thomas Edison, Henry Ford—big companies were synonymous with the individuals who founded them.

Then during the 1960s, as the business landscape came to be dominated by publicly traded companies without individual owners or prominent founders, modern management thinking came to see a company as a collection of distinct interacting functions. Senior management was seen as existing to increase the value of the company, in a financial sense—something logically distinct from the company's actual business. Managers became financial overseers, and involvement with actual products and markets became the province of corporate vice presidents, who often saw themselves as financial overseers of a market rather than visionaries of products.

The Internet age was like the Industrial Revolution in that it represented a pioneer period in a new landscape: the global network of the Internet. Just like the Industrial Revolution, huge companies grew, identified with their founders: Apple and Steve Jobs and Steve Wozniac; Google and Sergey Brin and Larry Page; Amazon and Jeff Bezos; Netflix and Reed Hastings.

But was this return to owner-dominated giant companies only due to the Internet? What about SpaceX, which was founded by Elon Musk? SpaceX has become a huge company, outdoing the likes of Boeing in the production of spacecraft, yet it was a tiny startup 10 years ago, and it had nothing to do with the Internet, although that is changing since they have undertaken to launch a network of Internet satellites. But the company's growth was based on its approach to building rockets—it was not due to the opportunities presented by the new Internet landscape.

What is noteworthy about these new companies is that their founders are deeply technical. In each case, the founder understood, or learned, the technology used to deliver the product or service. Steve Jobs and

Steve Wozniac were both home computer builders. Sergey Brin and Larry Page were both research scientists from Stanford. Jeff Bezos had degrees in electrical engineering and computer science from Princeton.

Elon Musk is an interesting case. He did not know much about rockets when he founded SpaceX, so he hired Tom Mueller, a rocket engineer. But Musk did not do what a modern management executive would do, which would be to concentrate on the business and delegate the product development to a VP. Instead, Musk said, "Teach me about rockets." Musk continued to drive the development of SpaceX's rockets, deferring to experts but always staying involved in the decisions—just as Steve Jobs personally approved of any product that Apple launched, and Jeff Bezos personally directed the development and technical parameters of Amazon Web Services.[3]

This personal involvement in the organization's products and services seems to be a common characteristic of managers of the most highly successful technology companies today. The view of these executives seems to be that the product delivery technology is not just a supporting function, subordinate to the company's strategy. Instead, it is *part* of the company's strategy.

How the company delivers is as important as *what* it delivers.

The value proposition of SpaceX is that it can launch satellites for less cost, and with higher frequency, than its competitors—substantially so. That is true only because of the *way* that it builds its rockets.

The value proposition of Amazon is that it can deliver products cheaper and faster than alternatives. That is true because of the *way* that its technology platform works. And Amazon can also update its platform rapidly and test market-facing features—and that is true only because of the *way* that it builds and delivers software to its online systems.

The *way*—the *how*—is not secondary. It is strategic. It is as important a differentiator as the product itself.

This means that today's executive cannot say, "That's a tech thing—my CTO deals with that." Today's executives need to understand how things are built, how they are delivered, how they work, and how they can go wrong. Elon Musk knows how rockets can fail because a failure can ruin the company's reputation. One of his main preoccupations is ensuring reliability. He achieves that by influencing how SpaceX builds the rockets. It uses what he calls a hardware-rich approach, whereby they don't try to perfect a design but instead perform early testing and iterate, constantly measuring and refining, until a design has proven itself to be robust and reliable.[4]

Again, the how matters—strategically.

Does this mean that one person needs to know how everything works and everyone's job function?

No, no one can know all that. But the person in charge of an area needs to know what they need to know—which things are important and which are not. To make that determination, they must have a vision for how everything works at a high level, and what can go wrong, so that they can zero in on those areas and make sure that the important issues are being decided the right way—and if they are not, to intervene and drive good decisions, either through spearheading discussions or setting guardrails.

Most of all, a leader cannot afford to say, "That is technical—I don't need to know about how the technical stuff is done." That won't work for a company whose business takes place on a technology platform.

What about everyone else? The CEO is not the only decision-maker in an organization. Yes, everyone else matters too. It is no longer the case that someone needs to know "only their job" the way that an assembly line worker at a Ford plant only needed to know how to attach a certain part. Today's organizations work in a more collaborative manner, so you need to understand what others do as well, to be able to participate in discussions with them and understand their needs and how your needs and their needs can be harmonized. Understanding the full range of activities (everyone else's jobs) also enables you to see opportunities for holistic improvement—approaches that make things better overall, instead of locally optimal.

It does not mean that everyone needs to be an expert in everything. That is not even possible. Instead, it means that everyone needs to take an interest in the entire value stream—the end-to-end sequence of activities that deliver value to customers or users. Learn as much as you can, instead of ignoring other job functions as "not your job."

By now you probably see a connection to Agile: we are advocating collaboration across job functions. One of the Agile Manifesto principles reads, "Business people and developers must work together daily throughout the project." The Agile Manifesto delineates between the business and the development teams.

There is actually nothing wrong with that statement. There are "businesspeople" and there are "developers" in most organizations, and they are separate job functions. However, the Agile community—as it has done with all of the manifesto—interprets it in the most extreme way. The business has been seen as something that feeds requirements—in the form of user stories—to development teams, usually via a backlog of work.

Thus, the business has been cast by Agile as apart from the developers—a dichotomy that is antithetical to the approach used by Elon Musk and Steve Jobs and Jeff Bezos and all the successful tech executives of our day, who know that there is no real separation between technology and business—that a successful product vision needs to be highly informed about the product, not just its features, but also how the product is built and delivered.

Tech Cannot Be an "Order Taker"

Agile treats a development team as a taker of orders: build this, and build that. That is the reality. In most Agile environments, the business and the developers are not partners. If products were like pizzas, where each is the same as the others before it, this order-taking approach would work, but when one designs and builds a new product, it is unique, and so a lot of creativity is needed. It is not mere execution. It requires inspiration and creativity. Taking orders does not inspire.

There is also an implicit assumption that innovation comes from the business, in the form of a product vision. That is how it is usually described. The developers are supposed to implement the work elements—the stories—to achieve the business's vision.

There is no mention of a vision coming from the technical side—the developers. In fact, the process of Scrum often—usually we would say—is a relentless mill of implementing one story after another. Many in the Agile community feel that Agile is often contorted into a kind of treadmill whereby teams are driven with the pressure of their sprint commitments, and the pressure never lets up, because they run from one sprint to the next, forever. Not only do many programmers feel that way, but one of us was told by an architect that she has observed that "Agile burns people out." Agile promotes a "sustainable pace" of work, but in many organizations it is a relentless pace, without peaks and lows.

One of us met an Agile coach who has worked in Silicon Valley since the 1990s. She said that the original sprints were followed by a period of rest. The choice of the word *sprint* was deliberate to describe a race to a finish line. Originally, there was never any intention of running race after race.

Developers have the weekend off—usually—but at work, the backlog is endless. It is a modern-day assembly line. There is no time for innovation or creativity among the programmers.

The pressure takes a toll as well. In 1908 psychologists Robert Yerkes and John Dodson documented the Yerkes-Dodson law, which showed that slight psychological pressure helps to drive performance but that increasing the pressure further decreases performance. In other words, relentless pressure on people does not make them deliver more or better work; it decreases their performance.[5] Psychological studies have also shown that pressure, or in fact anxiety of any kind, disrupts decision-making.[6]

Agile coaches will say that teams that are in that situation are not run properly, that the business—the Product Owner if they are using Scrum—needs to allow for periods in which programmers can experiment. That is a hard sell, though. Product Owners usually have little understanding of the technical work, so asking them to let the technical team experiment when they could be working on features that the Product Owner wants is like saying "Let them play around—your business features can wait."

The core problem is that Agile—in practice—is set up so that the business and the developers are not partners. The developers take orders from the business. That makes it unlikely that the developers will ever be able to offer an innovative approach.

But could developers have innovative ideas? If the ideas are technical, surely. Some of the most successful technology products—perhaps most truly game-changing technology products—were the inspiration of technical individuals. Examples include Linux, Skype, Napster, the first Apple computer, Microsoft's first operating system, Google, Amazon— these are but a few of a long list, and they changed or created new industries. A pure businessperson could not have thought of these things.

But what about in an established company, with a mature product? Will the developers think of innovative ideas pertaining to the product? Yes, *if they are familiar with how the product is being used*. That means that either they need to use the product themselves (programmers call that "eating their own dog food") or they need to talk to actual users. If they talk to actual users, they obtain an immensely better understanding of the product and how it needs to work. They become as insightful as the Product Owner, and they start to have their own innovative ideas—sometimes really fantastic ideas.

All too often Agile teams are not allowed to talk to actual users. They only talk to the Product Owner. The Agile Manifesto mentions the customer and the user, but not in the context of collaboration, and the reality of Agile development today is that development teams tend to operate apart from actual customers and real users.

This is a great dysfunction. Not only does it squander the creative ability of the technical side of the organization and convert them into an under-appreciated serfdom serving business stakeholders, but it isolates product marketing research and product vision into silos that the people creating the product have no view of—almost ensuring that the product will be suboptimal for its users.

Transformation Is a Journey, Not a Rollout

The traditional approach that organizations use for any large change is to create a big plan. Such a plan has tasks and milestone dates. The theory is that if you perform the tasks and they are done on time, then you will have completed the initiative. You will have "rolled out" the change.

When senior executives meet with their staff about the change initiative, they review a chart showing the status of each task, indicating whether they are on time, and the spend rate. If something slips, it is shown in red. A red task invites questions from the executive. Lots of red tasks indicate that the initiative is not going well—something that is not good for the career of the person leading the initiative. One wants to have all tasks shown in green.

There are several problems with this approach. One is that it discourages honesty. No one wants to report tasks that are in the red, so the tendency is to cover up problems. Scott Ambler has referred to this as *green shifting*. Such a cover-up is rationalized by the idea that "it will all work out," and so the progress of each individual task will not matter in the end, as long as the overall initiative succeeds. However, this means that problems are disguised, and so honest discussion—discussion that might help to ensure success—does not occur.

The ability to obtain people's honest beliefs and ideas is essential. You might think that things are fine, but they are not fine at all; and if people are not willing to be frank, they will not tell you what they think that you *should* be doing instead. As Jack Welch said in his book *Winning*,

> *"Lack of candor basically blocks smart ideas, fast action, and good people contributing all the stuff they've got."*[7]

Another problem is the assumption that the initiative is composed of tasks. If one is performing a largely repeatable process, such as building a new power plant using a design that has been used 50 times before, then one can indeed create a master task plan, and it will be helpful for managing the work. On the other hand, if the initiative is highly unique, then one cannot define all the tasks up front.

Another reason why a task model is not appropriate for a unique initiative is that for something unique, a large portion of the work consists of learning and trying things. A learning process is not well-defined as a task: it is ongoing. Learning occurs throughout the entire initiative, so one cannot, for example, define a task as "Learn how to perform deployment." This is because while one might learn how to perform a simple deployment, as the initiative progresses, deployments might become more complex, and one might learn better ways and continue to refine the approach. Thus, the learning process is continuous and is never finished.

Agile transformations are extremely driven by learning. This is because most of Agile is not about *what* you do—it is about *how* you do it. Agile advocates judgment, instead of prescribed steps. To learn judgment, you must do something again and again over time. That kind of learning is not neatly confined to a task, yet the transformation is not complete until learning has completed—which is, well, never. That is why people say that a transformation is not a destination; it is a journey. You never get there, but you get better and better. Your metrics become better and better. Along the way, you learn that some approaches are not working out, and you change direction.

Some tasks definitely can be defined. But a task model is the wrong approach for planning the transformation and for measuring progress. Instead, what works better is to have various ongoing and intersecting initiatives for problem identification, for upskilling, for collaborating about approaches, for creating and refining metrics, and for instituting and refining various practices.

Unfortunately, there is information on this topic that is easily misunderstood. The person who first connected the dots about DevOps, Jez Humble, gave some brilliant talks during the decade of the 2000s and came out with his groundbreaking book *Continuous Delivery* (co-authored with Dave Farley) in 2011. But in 2018 he co-authored (with others) a book called *Accelerate*, which is a great book, based on a collaboration with Dr. Nicole Forsgren. The book addressed the problem of transformation: how to help an organization progress from a current state to one that uses DevOps methods. All good so far.

Dr. Forsgren's work is based on a capability model, and if you read the book, the capabilities are intelligently defined. In fact, they are what Agilists would call *practices*: they are about *how* you do things. But Dr. Forsgren calls them *capabilities* and, by doing so, implicitly (perhaps unintentionally) links them to capability models that are common for old-style IT change initiatives.

These old-style change initiatives were invariably task-centric, and the capabilities usually consisted of the "rollout" of new tools.

That is not what Dr. Forsgren's work is about. It is really good work—perhaps great work—and the book *Accelerate* is a great book. But the use of the term *capability model* has the potential to make executives think, "Okay, I know capability models. I'll just create one, and we're done." And what they often create—we know, because we have seen this happen first-hand—is an old-style capability model that defines a set of DevOps tools, as well as nonsensical capabilities such as testing. We say nonsensical because testing is 90% of DevOps, and what matters is not that you have a testing capability but that you are doing it the *right way*.

Our point is, don't treat transformation as a rollout or as a process that you can define ahead of time, like a manufacturing process. And don't use a capability model unless you carefully read *Accelerate* and define your capabilities *their* way.

The Individual Matters As Much As the Team

The team, the team, the team!

Agilists are obsessed with the team. The word *team* comes up in almost every sentence. What about the team *members?* People are individuals. Yes, the team matters, but so do individuals, and the importance of the individual seems to have been lost in the Agile community.

We find this ironic because the first value of the four values of the Agile Manifesto begins with "Individuals." It does not begin with "Teams."

Yet today, one never hears about the individual in Agile circles; it is always "the team." This is also confounding because today's products usually cannot be built by a team; they require *many* teams. So really, people should talk about "the teams," but they rarely do; it is always the team, singular, as if each team exists unto only itself.

There is a maxim in the Agile community: "autonomous self-organizing teams." However, as we will see later in this book, there is rarely such a thing; more common is *somewhat* autonomous, *somewhat* self-organizing teams. So teams do not usually operate independently—independence is actually a worthy goal, but it is rarely achieved—and so a focus entirely on the team, singular, is not realistic.

The worse problem, though, is the loss of acknowledgment of individuality. The culture of the Agile community is so biased toward the team that being different from the team is seen as being a misfit. We have read blog posts in which experienced Agile coaches say that if someone works differently than others on the team, then the team can do without them.

The team ethos is so extreme that it is sometimes compared to a communistic view; the individual is entirely subordinated. Agile coaches discourage any kind of individual recognition; only a team should be recognized for its success. Different levels of career experience are generally not seen as significant, despite that experience does make a world of difference, which we will discuss later. The egalitarian view that anyone can work on anything permeates Agile culture, yet the reality is usually very different.

People are individuals, they have careers, and they have financial pressures and personal needs. They want to advance in their careers, so the idea that everyone is equal translates into no one can advance: if you are on a team, you will always be on a team, and you will never progress in your career. Your income will never increase, because the value of your experience will not be acknowledged. When you become expert in working on certain things, the value of that will not be recognized because, according to Agile theory, anyone can work on anything.

Fortunately, real companies do not usually operate this way. Most real companies have pay grades and career levels. But the Agile community is a parallel universe. Levels and individual differences do not fit the Agile narrative, so there is no framework for discussing it in an Agile context.

Teams matter. Teams are powerful. Teams are a valid construct for developing software. Extremes are what are bad. The extreme idea that only a team matters is not realistic and is unfair. The extreme idea that no one can advance is unfair. The extreme idea that anyone can work on anything has the noble goal of empowerment and opportunity for all, but by itself it is extreme and must be balanced with the reality that there are experts too—more on that later.

Culture: Individual vs. the Collective

There is a cultural struggle in human societies pertaining to focus on the individual versus the group—the *collective*. It might stem from a difference of values, in terms of whether a culture generally prioritizes equality over individual liberty (freedom) or individual liberty over equality.

During the mid-1900s, Jean-Paul Sartre and Albert Camus were fast friends. Both were socialists, as were most people among the intellectual community at that time in France. However, over time the two philosophers realized that they disagreed at a fundamental level: while both valued equality and liberty, Camus found that he valued liberty more than equality, while Sartre realized that he valued equality more than liberty.

The two could not reconcile their difference of opinion, because it was deeply rooted in their respective values; and since their opinions defined them as philosophers, they had a famous falling out.[8]

Most modern cultures place a high value on equality and the importance of the group, and also most place a high value on individual liberty. However, some value one over the other.

For example, North American culture tends to value individual freedom and liberty more than equality. The importance of liberty traces to the founding of the Americas by people who were searching for a land where they could practice their religion their own way. The subsequent settling of the Americas by pioneers reinforced the importance of the hardy individual and self-reliance.

In contrast, many European cultures place a higher value on the community than the individual. Perhaps that is why so many European countries have strong social safety nets, and in many European countries, it is commonplace for union representatives to sit on corporate boards.

Central and South Asian countries have cultural patterns as well pertaining to individualism and the group. In many Southeast Asian cultures, the group is valued more than the individual, and age and seniority are considered extremely important. These values are in conflict to some extent with the tradition of Agile, which advocates an "everyone is equal" and "anyone can work on anything" mindset.

Which is better? Is it better to value the individual or the group? It depends on your value system. It is also impossible to say, in a strict technical sense.

Visionaries usually call upon us to follow them on a path that is atypical and that experts advise against. That requires a great deal of trust, and that is why visionaries need a source of influence: either they have a powerful and persuasive personality or they have great resources in their name.

Is it better to let visionaries dictate our path or to block would-be visionaries and always compel their subordination to a group of experts who presumably know better?

Given the emphasis on the individual that one finds in North America, it is no surprise that North America has so many startup companies. In North America, the individual is the star: Jeff Bezos, Steve Jobs, Elon Musk, Larry Page and Sergey Brin, Henry Ford, Thomas Edison. The list goes on and on—they are visionaries who are celebrated.

But what about Elizabeth Holmes? What about Bernie Madoff? Michael Milken? Kenneth Lay?—antiheroes who could have been heroes if their efforts had panned out in a positive way.

How does one know that a person is a visionary? Can one know ahead of time? How does one know if one should trust the experts or trust the visionary?

There is no way to know. Most wannabe visionaries are often wrong, perhaps usually wrong. However, it is visionaries who change the paradigm. Given enough time, the paradigm would probably shift, but visionaries make it shift now. It was visionaries who gave us the iPhone and its copycats, electric cars, and relativity (which was stridently mocked and refuted by many scientists of the time).

So, do you follow a visionary or follow the group consensus? There is no way to know, and it also depends on your value system.

People Don't All Work the Same

Individuality also pertains to *how people work*. Not everyone works the same. Just because someone works differently than others on a team does not make that person a misfit. Agile is rife with all-team practices: everyone joins the standup; everyone is equal in the retrospective; everyone takes a story and works on it, collaborating with others. But some people do not adjust well to those ways of working—ways that are highly interactive, in which you are on the spot all the time, on your feet or expressing your idea verbally. Some people work better in isolation, with quiet, and express themselves better in writing.

Such people are not outliers; they are the introverts, as we discussed in Chapter 1. If we shun those people or make them feel like they are misfits, we lose the enormous value of what they can contribute. Those people are often the deep thinkers—the ones with the powerful insights. It was during quiet reflection before a fireplace that Dirac suddenly realized the solution that became the Dirac equation—the key to relativistic quantum mechanics.[9] It was a moment alone, making his morning coffee, in which Héctor A. Chaparro-Romo suddenly saw the solution to the 2,000-year-old problem of spherical aberration.[10] It was a walk along the Royal Canal in Dublin when Sir William Rowan Hamilton suddenly saw the solution to the quaternion problem,[11] which underlies much of particle physics today.

Yes, these people collaborated with others, which led to new trains of thought, which is essential; but it was during silent reflection that they were able to create the deep and complex mental model needed to pursue those new trains of thought and suddenly have the insight needed to solve a complex problem.

It is a shame—a poverty—to ostracize such people and make them feel like misfits on teams.

Communication Is a Process, Not an Event

We have mentioned that before the Agile Manifesto was created, Alistair Cockburn had been writing about how, in his opinion, face-to-face communication is the most effective form of communication. It is not always, though.

Communication about complex topics is a process, not an event. If you gather five people into a room for an hour to talk about something complicated, which none of them have had a chance to discuss together before, then what usually happens is that in the course of the hour the conversation will jump all over the place. Some will be unable to fully explain their thoughts. They can't because for a complex topic, it might take longer than people are willing to listen without jumping in.

The situation is different if they have all been immersed in the subject. In that case, they have a shared understanding as a baseline, and they can discuss incremental ideas relative to that. But if someone is far ahead of the others, they won't be able to lay out their thoughts and not have someone inject tangential issues that will take up the rest of the hour.

In the realm of IT, there is a common topic that is complex. It pertains to branch/merge strategy. The issue is even more complex if one is considering multiple code repositories. The various issues pertaining to this are way too complex to discuss effectively in person, unless there is a shared baseline understanding of the issues. That is why one of us has used the approach of writing up a white paper, explaining the issues and laying out options, distributing it, and *then* having a meeting to discuss it.

Similarly, the idea that after a challenging product launch a team can gather to talk through what happened, what went wrong, and how to do things differently next time is simplistic. One of us has the approach of spending a week inviting team members to record and share their ideas in writing anonymously, summarizing the input and sending that out, then asking the group to submit solutions or action items in writing anonymously, and only at the end of the process coming together in person to discuss and prioritize what to do next. This gives people time to think, to write, to reflect, and to share ideas asynchronously over time, before having a meeting to discuss it. Indeed, we took a similar approach to developing the ideas of Agile 2.

The point is that the idea that face-to-face communication is always best is—like so much of the Agile community's interpretation of the Agile Manifesto—an oversimplification of real-life situations. Unfortunately, it has led to a community bias against writing down one's ideas. Writing is sometimes essential. What is writing for, if we are not supposed to do it?

Jeff Bezos famously makes his staff write a "six-page, narratively structured memo" for each meeting, and he says that it was the "smartest thing we ever did" at Amazon.[12] He basically wants his people to think deeply and not have shallow debates in which they go in circles. He believes that complex issues require structure, and the best way to establish structure is through writing.

Our human brains are not like computers—they are not designed to be information storage vehicles. In *The Organized Mind*, Daniel Levithin explains that memory is fickle, and while we're good at making snap life-or-death decisions, we're not good at storage—thus the first forms of writing were lists—humans offloading the storage to free up brain space for more complex decision-making. Rather than supplant face-to-face communication, we believe that well-written, clear documentation enhances communication. By writing down a structure, the face-to-face conversations become more focused and more effective.

Communication is a process—not an event—and face-to-face is not always the best way to start. Sometimes a combination of communication formats is most effective, and the person who is organizing the discussion needs to think about what communication methods, in what order, might work best for the issue and also consider the innate communication styles of the participants, since some people communicate differently than others, as we have already learned.

The Importance of Focus

According to an article in *Inc*, groups that collaborate less often may be better at problem solving.[13] The article argues that the always-on environment created by tools like Slack, desktop pop-up notifications, and sitting within earshot of others actually disrupts one's ability to focus. The article cites research that shows that intermittent collaboration—that is, collaboration that follows periods in which one has no interruptions—leads to the best outcomes.

The noise of an open office also has been shown to reduce cognitive performance.[14] The *New Yorker* examined the impact of its new open plan workplace and determined that "the environment ultimately damages workers' attention spans, productivity, creative thinking, and satisfaction. Furthermore, a sense of privacy boosts job performance, while the opposite can cause feelings of helplessness."[15]

In other words, *the Agile team room is dumbing us down.*

The article differentiates between several situations in which collaboration is needed.

- Planning
- Ongoing improvements
- Consensus or decision about an issue
- A crisis

In each of these situations, a group needs to synthesize its ideas and ultimately act as one. However, note that communication is not mentioned. Communication about what someone is doing, or challenges they are having, does not necessarily need immediate discussion, so why bring a whole group together for it? It might be better to provide the information in a real-time dashboard or a group message. And if the message does not interrupt but can be ready when each person is ready—when they need a break from their work—then it does not interfere with their focus.

One of the standard Agile communication practices is the daily standup. A standup meeting is supposed to be a short meeting—usually "time-boxed" to 15 minutes. In a standup, the team members all stand, and a facilitator goes around the room, asking each person what they have accomplished, what they plan to work on today, and what their "impediments" are.

Many people in the Agile community say that a standup is not a status meeting, but it is hard to imagine a meeting that more accurately defines a status meeting than "what did you do, what will you do, and what is in your way?"—that set of questions epitomizes a status meeting.

The problem is, it wastes everyone's time. Not everyone wants—or needs—to know what everyone is working on.

In Agile 2 there is a principle that goes, "The whole team solves the whole problem." This means everyone on a team needs to understand how things will fit together. They all participated in sketching out the solution, but from that point on, they don't need to know the details of each other's progress. They only need to know about issues that arise that might affect them, and they definitely need to know if there are any changes in how the overall solution will work.

We hear so many programmers say things like, "When I hear people in a standup say what they are working on, I don't understand most of it, and I don't really care what they are working on. I only want to know about what affects me."

We need to listen to that, because it's the people who are doing the work telling us, and a core Agile principle has always been to let people decide how they want to work.

Some Agilists will immediately respond that if team members are not interested in what others are working on, that is a sign of other problems. It might be, but it might not be. Once a team has decided how it is going to implement a set of features, most of the team members might not need to communicate very much.

What product teams *do* want to know about is anything that changes pertaining to how all the different parts of the solution fit together. And they want to know if something changes that directly affects what they are working on, including new information about how the customer is using the features that they are working on or have worked on. They usually don't want to know any more than that!

So, we should stop demanding that they give up part of their morning to attend a standup. You might think, well, it is only 15 minutes. It is more than that, though. An article in *Business Matters* magazine quotes John Jenson, technical director at TandemSeven: "The stops and starts introduce mental context switches for all involved. If we apply a 15 min tax for each developer for each meeting, we lose another 30 hours."[16]

But actually it is *a lot* more than that. When someone knows that they are going to be interrupted, they don't start anything that requires them to think deeply. So if a product developer arrives at work at 9:30 a.m. (a not atypical start time for technical people) and there is a standup approaching at 11 a.m., they will check their email and respond to Slack messages and get their coffee and then be ready to dive into work by 10; but they are then going to be leery of "getting deep into anything" before the standup, because they know that they will not be able to get far before being interrupted by the standup. After the standup, which often runs over, it will be approaching 11:30 a.m., which is close to lunchtime, so again, they are not going to want to dive deep into their work; they will stay at a surface level and then break for lunch. By the way, a standup is a group meeting, which is taxing for an introvert, and so the introverts will likely need a little decompression time after the standup before they can focus again.

So, a standup is actually *very* costly. Is it worth it? Ask your team what collaboration formats will work best for them at this particular point in the lifecycle of their work. Listen to them.

Also remember that the standup might be a way for the team lead, if there is one, to learn what is going on; however, there might be other ways to do that, such as by checking in briefly with each person individually. That makes richer discussion possible if it is actually needed. Some teams use a tool that enables them to broadcast a concise summary to their team of what they are doing and if they have any issues. In the words of one Agile coach,

> "A tool I recently starting using is called Status Hero. I highly recommend it. My favorite feature — it prompts your team to include an emoji about how they are feeling. You can bet I am interested in the feelings of my team."[17]

Remember that focus is just as important as collaboration: if you destroy focus, your product will be terrible.

Data Is Strategic

The Agile Manifesto said nothing about data. This oversight is really quite amazing given that data has always been considered to be a strategic asset. Also, data is the counterpart of a working product, and so it is therefore reasonable that practices pertaining to data are different from but relevant to practices pertaining to product development.

There are at least four ways in which data must be considered, apart from one's product.

- **Historical data** about one's customers, constituents, users, and so on, is a rich source from which insights can be derived if—and only if—the data is managed so that it can be understood, correlated, and analyzed after the fact.
- **Transactional data about current operations** is essential for product developers to model and understand so that they can correctly extend the products and services.
- **Production-like test data** is essential for product developers to be able to validate one's product. Too often product development teams are left to cobble together their own test data, when business stakeholders own the production data and are in the best position to do it. Yet the business stakeholders expect a working product.
- **Protection of data** about customers, constituents, or any individual or organization is potentially sensitive and needs to be safeguarded. This is often an afterthought, even though the security community has been proclaiming loudly for years that the only way to secure data is to have product developers build security in from the beginning.

These are all major gaps in today's common practices, and Agile says little about any of them.

The DevSecOps movement arose during the mid-2010s, tacking security onto the DevOps moniker, but it usually represents no more than automated scanning tools. For some reason, the industry has almost zero interest in doing what works: make product developers get certified in secure product development, even though there is plenty of training available for that.

If you think that data security or privacy has nothing to do with Agile, let us tell you how wrong you are. In the early 2000s when the Agile movement began, people who were building large-scale enterprise systems viewed it with interest but were dismayed by the lack of mention of anything about security and reliability. The Agile community took a cue from the lack of mention of security or reliability by the Agile Manifesto, and the community displayed little or no interest in security and reliability from the beginning. But anyone who builds enterprise-class systems knows how foolish that is.

Security and reliability need to be designed in. Here is an illustration. One of us has an interview practice, in which he asks a product developer to sketch out a design for a system. The candidate draws a diagram on a whiteboard. The interviewer then says, "Now, suppose it has to be very secure." The candidate looks at their diagram, thinks for a bit, and then invariably creates an entirely new diagram.

The point is, if Agile is to inform us about how products and services should be built, it needs to emphasize that those products and services contain sensitive assets—data—and that protecting that needs to be a core value or principle. To fail to even mention that is akin to the Ten Commandments failing to mention "Thou shalt not kill."

What about the other ways in which data should be considered? For example, the role of historical data? Too often we see microservice teams dump data into a data lake without documenting the structure of the data. A machine learning colleague of one of us complained that he was hired by a Fortune 10 company to create a system that could be trained from their data lake's data, but he was not able to match up data from different sources in the data lake: there was no schema that bridged the various sources—no "map" or "translation" linking the various data domains. Worse, many teams use NoSQL databases and fail to document the data structures that they store. The data lake was essentially a wasteland of unintelligible data.

As a result, this company has a huge asset—its customer, product, and sales data—and is unable to make any use of the data.

What about transactional data? One of the software methodologies that is often cited as an Agile method is Feature-Driven Development (FDD). In this method, the first step is to create a model of the organization's information. That model is then used by product teams as a shared understanding of the transactional data. Other methodologies that claim to be Agile do not mention data; for example, Scrum says nothing about it and Extreme Programming mentions it in passing. How can it be that one methodology (FDD) is built around something that the others do not even mention?

What happens with Scrum teams is that when they have to use existing transactional systems, the team members take on the task of learning about those systems. To do that, they need to request help from other teams that are familiar with those systems. Thus, the process of sharing knowledge about the organization's information is entirely ad hoc. Is that a good thing? For something as fundamental as the data that is being used by all of the organization's systems, does it not make sense to establish that there is a universal shared view of that information? Be your own judge.

Finally, the issue of test data is a crisis-level one. One of us was in a program-level meeting and heard one development lead say to another, "My tests always pass because I create data that I know will pass," and they both laughed. But as absurd as it sounds, it is actually often the case. Product developers create test cases and test data, but they do so based on their understanding of things—an unvalidated understanding. Those tests are therefore not valid with respect to the actual structure of the organization's data. The only way to verify if things will work with product data is to test with production-like data.

Yet what usually happens is that development teams are told to create a new set of features and obtain their own test data. How? The production data is usually owned by a business function; so the developers go and talk to the people who support the systems, who have access to the production data, and say, "Can you get us some production data?" That is a very ad hoc approach and is a recipe for problems when the new features go live.

Some organizations have test accounts set up so that development teams can access production systems using test accounts and thereby access production-like data. Organizations also sometimes "mask" or "surrogate" production data to hide sensitive data from development teams. Those are effective practices. The bottom line is that data for testing is a critical need: if development teams do not have ready access, it is the responsibility of the business function that owns the data to help the development teams solve that problem and obtain production-like data.

Notes

1. agileforest.com/2012/02/26/scrum-evolution-over-time-part-2-roles/
2. www.forbes.com/sites/cognitiveworld/2019/08/23/the-end-of-agile/
3. apievangelist.com/2012/01/12/the-secret-to-amazons-success-internal-apis/
4. Elon Musk on how SpaceX designs and tests: www.youtube.com/watch?v=xNqs_S-zEBY
5. en.wikipedia.org/wiki/Yerkes%E2%80%93Dodson_law
6. medicalxpress.com/news/2016-03-bad-decision-anxiety-blame.html
7. Jack Welch; Suzy Welch. *Winning*. HarperCollins e-books. Kindle Edition (p. 25).
8. press.uchicago.edu/Misc/Chicago/027961.html
9. en.wikipedia.org/wiki/Dirac_equation#Dirac's_coup
10. petapixel.com/2019/07/05/goodbye-aberration-physicist-solves-2000-year-old-optical-problem/
11. en.wikipedia.org/wiki/History_of_quaternions#Hamilton's_discovery
12. www.cnbc.com/2019/10/14/jeff-bezos-this-is-the-smartest-thing-we-ever-did-at-amazon.html
13. www.inc.com/tanya-hall/all-that-collaboration-is-hurting-your-results-heres-why.html
14. www.sciencedirect.com/science/article/abs/pii/S0272494408000728
15. www.washingtonpost.com/posteverything/wp/2014/12/30/google-got-it-wrong-the-open-office-trend-is-destroying-the-workplace/
16. www.bmmagazine.co.uk/in-business/not-all-developers-like-agile-and-here-are-5-reasons-why/
17. blog.echobind.com/4-ways-to-optimize-an-agile-environment-with-nudge-management-2adbb2878d9f

3 Leadership: The Core Issue

If an organization has good leadership, everything else will follow. If an organization has bad leadership, no methodology or set of rules will save you.

Elon Musk's companies have demonstrated not only unprecedented innovation but competitiveness, altering and taking the lead in entire industries. Musk sees innovation as a strategic process to be managed and grown. When Robert Zubrin, president of the Mars Society, asked Musk what it will take to get to Mars, Musk's response was not about rockets or money; his response was about the needed rate of innovation:

> *"I'm trying to make sure that our rate of innovation increases. . . this is really essential. . . if we do not see something close to an exponential improvement in our rate of innovation we will not reach Mars."[1]*

Zubrin then asked about Musk's methodology. Zubrin said this:

> *"One thing that is really amazing about SpaceX to those of us who have experience in the aerospace industry is the rate of innovation. Last time you spoke to the Mars Society convention it was 2012. Since then you have made Falcon 9 reusable, introduced Falcon Heavy, Crew Dragon, a satellite constellation, and you're in the middle of developing Starship. What is your methodology that allows you to innovate so swiftly?"*

Musk's response was, after a long pause, "I don't really know."

What makes SpaceX and Musk's other companies so successful and so innovative is not a methodology; it is the leadership that Musk provides.

This is why leadership is the core issue for understanding how a group of individuals can work together effectively. The issue cannot be sidestepped. Bad leadership is often a terrible problem, but dismissing the need for leadership does not solve the problem. We must deal with it head-on.

Authority Is Sometimes Necessary

The Agile Manifesto tried to address the issue of the bad manager by ignoring the need for managers, implicitly saying that managers are not part of the equation. That is like trying to deal with bad friendships by dispensing with friendship. According to Mark Schwartz, former CIO of the US Citizenship and Immigration Service and a well-known Agile and DevOps evangelist,

> *"The Agile world, ever suspicious of management, proceeds as if it can manage without the involvement of IT leaders."*[2]

A manager is, by definition, a leader who has authority. Authority at the top is unavoidable. An organization has owners or shareholders and a board or a government-appointed leader who has oversight. The question is, is authority needed at other levels?

The Agile Manifesto is silent about the role of managers. Typically, in most organizations, a manager has direct staff, and those people are the manager's team. The Agile Manifesto's principle "The best architectures, requirements, and designs emerge from self-organizing teams" seems to advocate that teams do best when self-organized, which implies that managers play no direct role. The Agile community has struggled over the years to figure out how to integrate managers into Agile ways of working.

While the Agile Manifesto was written in the context of software development, the ideas have been applied elsewhere, and the culture of the Agile community strongly reflects the self-organization ethos, regardless of the domain of application.

Yet if successful Internet companies can be a guide, managers and team leads are still very much present at the most successful companies. For example, at Google most development teams have team leads. According to the book *Software Engineering at Google,*

> *"Whereas every engineering team generally has a leader, they acquire those leaders in different ways. This is certainly true at Google; sometimes an experienced manager comes in to run a team, and sometimes an individual contributor is promoted into a leadership position (usually of a smaller team)."[3]*

The challenge with collective leadership is that authority is sometimes needed, and while a team can collectively have authority, authority requires accountability, and it is difficult to hold a whole team accountable.

The Agile community is right to have anxiety about authority. Traditional organizations use authority way too much. The traditional Theory X model of a manager who dictates how work should be done and expects everyone to follow orders might work fine in some situations, but most of the time that approach works very poorly, particularly when judgment and creativity are important components of the work. We will discuss Theory X and other leadership models in the "Theory X, Theory Y, and Mission Command" section later in this chapter.

Even when work is repetitive and uncreative, allowing people some control over how they do the work leverages their experience with the tasks and also gives them an important feeling of personal control, which boosts morale.

Authority is needed, but it should be used sparingly. *Having authority does not mean that you use it.* In fact, people often conflate two ideas: (1) autonomy that has been granted and (2) no one having authority. These are not the same. Authority may be needed to cover many situations, but the best use of authority is often to give others a reasonable degree of autonomy.

Often the use of authority, especially in the form of micromanagement, is not needed. If you dictate what people should do and how they do it, you fail to leverage their ideas and their experience, and you make them feel disempowered. No one wants to be just an order taker.

On the other hand, a leader sometimes needs to make a final decision. A great example of this is Elon Musk's decision to have Tesla develop its own battery technology in-house. An article in *Teslarati* chronicles this, and Elon Musk's role in that development. According to the article,

> *"Musk's subordinates have reportedly argued against the idea of developing proprietary battery cells, but the CEO has been adamant about his goal."[4]*

The CEO (Musk) made the final decision, and today Tesla's battery technology is changing the industry.

Musk could have been wrong, though. There is no way to ensure that the person or group with the best judgment about a particular issue will get to make the decision about it. Quibi was supposed to revolutionize Hollywood by bringing movies to mobile devices, but the instincts and vision of Jeffrey Katzenberg and Meg Whitman—two highly credible industry insiders—proved wrong.[5]

There is no fail-safe approach. Leadership should try to make sure that those who have the most experience, depth of knowledge, insight, and vision about an issue are all able to consider and discuss it openly in a manner that encourages everyone to contribute to the discussion, and that those who have the most invested in any sense will have the final say—informed by everyone else's thoughts.

The Path-Goal Leadership Model

As we said, having authority does not mean that you always use it. A leadership model known as Path-Goal Theory[6] posits four leadership styles.

- Directive
- Achievement oriented
- Participative
- Supportive

A directive leader is one who issues commands and expects others to follow. An achievement-oriented leader is hands-off but sets objectives for team members. A participative leader is collaborative and engages in group decision-making. A supportive leader is more focused on the well-being of the members than on the business objective.

Few people fit these patterns perfectly, of course. This is just a model, but it is useful. We introduce the terms here to illustrate different styles of leadership and to set the stage with the idea that there are many forms of leadership. Please bear that in mind when reading the following sections. We will consider forms of leadership in more detail later in the chapter.

We also want to state that no one form of leadership is better than the others. They each have their place, depending on the situation, and sometimes more than one style is needed.

Collective Governance Does Not Solve the Problem

People have tried to figure out if organizations can be structured in a way that authority can be bypassed. Perhaps if there are governance rules, then everyone can be equal, and the system will become collectively governed. Everyone has a vote, in a sense. The most well-known approach for that is the holacracy model.

Does it work? It can, if you hire just the right people. But no one has shown that the model is easily repeatable. And it is not clear that it actually works that well. Medium tried it and abandoned it. According to Jennifer Reingold, writing in *Forbes* about Medium's experience,

> *"The sheer number of rules and regulations, combined with the potential for politics to seep in in different forms, makes holacracy, in my view, a questionable replacement for the classic management system, as flawed as the latter may be."*[7]

What about giving up on governance altogether and just letting people self-organize? Let leaders emerge—whoever they may be.

It seems that a lack of structure actually leads to hidden power structures that are as stifling—or more so—as formal structures can be. An article in *Wired* relates the story of Jo Freeman, a 1960s women's liberation movement icon. She complains that "If anything, the lack of structure made the situation worse." That is, the hidden power of male-dominated systems was more oppressive than entrenched but explicit power structures.[8]

Perhaps authority structures are needed, but simply fewer of them. This leads to the idea of "flat" organizations—ones that have fewer managers and therefore fewer levels of hierarchy. Google famously tried that approach in 2002, but it did not last long. According to an article in *Fast Company*, "Folks were coming to Larry Page with questions about expense reports and interpersonal conflicts."[9]

Perhaps the preference for self-organization is cultural. One of the members of the Agile 2 team claimed that the Agile community's beliefs about leadership and team behavior reflect a Silicon Valley perspective. Another pointed out that Silicon Valley culture tends to value, in his words, "innovation, freedom, entrepreneurship, collaboration, shared ownership, and anarchism." Those values seem fairly well-aligned with Agile values and attitudes as they are typically expressed.

Does this make Agile incompatible with some human cultures? We are not sure; but to us, the conclusion must be that people need to be able to be selective about Agile ideas and apply them in their own way, rather than using a one-size-fits-all set of practices.

There is an inherent belief in Western society that democracy is a fair process; and yet, democracy can lead to "tyranny of the majority," whereby the majority vote to subjugate a particular minority. Democracy is not inherently fair. German sociologist Robert Michels proposed the "iron law of oligarchy," which posited that any democratic system will inevitably devolve into an elite oligarchy.[10] Thus, the assumption that even a completely egalitarian team operates in a fair manner cannot be assumed to be true.

Another problem is that in any self-governed system, leaders emerge; informal authority develops. And so even though there are no *appointed* leaders, there are still leaders. No one appointed Genghis Khan to lead the many tribes of Mongolia to create the Mongolian empire: he appointed himself, through the influence and power that he developed. No one appointed Augustus to lead the Roman empire.

In any group, leaders usually emerge, irrespective of any governance structure. In the United States, Donald Trump was not even a politician but managed to develop enough influence to get elected to be a US president. Influence is leadership, and influence creates its own kind of de facto authority, and it is not always good leadership. Adolf Hitler became a world leader not through peaceful ascent through the ranks but through party politics and the popularity and therefore informal influence that he developed through his speeches, writing, and acts of intimidation, with disastrous results. Leaders can, with effort, undermine any governance structure. Russia has a Constitution, but its leader seems able to do whatever he chooses. Rules of behavior do not ensure that a team will behave well or that there will not be a lot of unfairness going on.

A study published in the *Journal of Business and Psychology* has shown that in-person teams tend to choose leaders who are confident, magnetic, smart-seeming, and extroverted; but in the study, remote teams chose leaders who "were doers, who tended towards planning, connecting teammates with help and resources, keeping an eye on upcoming tasks and, most importantly, getting things done. These leaders were goal-focused, productive, dependable and helpful."[11]

The implication is that in-person teams do not necessarily choose good leaders but instead choose leaders who "look good"—who look like leaders but might actually be poor ones. This indicates that one should not trust or rely on emergent leadership—at least not for in-person teams.

Self-organization also assumes that a team will eventually learn to avoid or resolve conflict. However, conflict can easily tear a team apart. Organizational psychologist Marta Wilson writes about the dangers of conflict if it is allowed to persist.

"Once disagreement takes on a life of its own, the chance that resolution will arise simply out of the group dynamic decreases, and a real conflict begins to gain a foothold."[12]

Having leaders available to help teams to resolve conflicts is essential.[13] Such teams are therefore not entirely self-organizing, but are *mostly* self-organizing.

Conflict is a natural result of differing opinions being discussed. It is an outcome of group creativity.[14] This means that if a team avoids conflict, it might be doing so by stifling discussion of important topics. It is therefore important to view a moderate level of conflict as natural and valuable, as long as people behave respectfully. Avoiding conflict is not a good strategy, but since conflict can lead to team disruption, having external leaders available who can help to stabilize things is important.

Leadership is a critically important issue, because if you have good leadership, the methodology that your teams use will matter less; things will likely go well. Conversely, if you have bad leadership, things will not go well—no matter what methodology your teams use.

This means that the questions of leadership and authority cannot be bypassed. We cannot remove leaders from the equation. Leadership and authority—formal or informal—will always exist in any collection of people, and we must take it into account and not wish it away.

Dimensions, Modes, Forms, and Directions of Leadership

Leadership is *influence*: a person is a leader if they have influence over others. This includes influence of any kind.

You might have influence over a friend, but you don't have actual authority over them. Your influence might lie in their respect for your ideas or your enthusiasm, which is catching. Whatever the source, that kind of influence does not come from having explicit authority.

Thought leaders usually have no direct authority over others. People read Deepak Chopra's books because they respect what he says. In that way, he influences them; he leads them. His readers have agency. They do not follow Chopra on command, but they do follow him—willingly—because they believe that he has insight. People often follow others because they feel those others have insight. One can follow someone even without fully understanding the other, if one trusts the insight and judgment of the other.

What about the Path-Goal leadership model? In those terms, we might say that Chopra's leadership style is both achievement oriented and supportive. It is achievement oriented, because he describes ideas for people to embrace, thereby challenging them; and his style is also supportive, because his advice pertains to *their* well-being, rather than to achieving Chopra's own objectives.

Chopra is a thought leader. On a team, there are often people who develop influence through thought leadership. They might have a great deal of experience or they might have deep knowledge about a topic or they might have shown that they are very smart or think things through well.

Others on a team might develop influence through their force of personality: they are persuasive, or they project an air of authority. They look, sound, or act "like a leader."

It is also not uncommon for cliques to emerge among groups of people, including work teams. A clique can be a source of power: if one attains influence within a clique, the clique can exert collective influence over the entire group, through the force of its numbers. If the clique has a leader—which is often the case—then that leader, in effect, can control the entire group, becoming, in effect, a directive leader.

This clique control pattern is described by Leader Member Exchange (LMX) theory,[15] which is primarily a descriptive theory rather than an explanation of why people behave that way. LMX theory observes that "in-groups" or cliques naturally form, and therefore intentional intervention is needed to prevent them or dismantle them. Doing so requires either authority or explicit preventions such as group voting, but even with such preventions, people still develop influence; thus, we are back to the power of informal influence and its ability to circumvent any procedural safeguard.

It is important to note that any type of influence is a kind of authority; it is a degree of authority that others give to someone over themselves. They follow the influencer willingly, but they still follow. This is informal authority, and just as formal (explicitly given) authority can be taken away by whoever granted it, informal authority can be taken away by those who follow.

A leader can delegate to others; they thereby lend their authority to those others—whether the authority is formal or informal. A democratic vote is a mechanism by which a group delegates their collective authority to a single member, nominating the person as their leader. That leader then might operate in a directive, achievement-oriented, participative, or supportive manner.

Authority can also be constrained. For example, one might have authority over whether a team can release a product to the users, but one might not have any other kind of authority. As another example, one might have authority over product feature decisions within a team, but no other kind of authority.

Soft Forms of Leadership

We mentioned Deepak Chopra as a thought leader, which is a kind of "soft" leadership—that is, leadership that does not work through making commands. It is not directive. Deepak is not telling anyone what to do; rather, people read his advice and incorporate it into their thinking. In that way, they reflect and learn and ideally grow in their understanding of things. There are some specific modes of soft leadership: *coaching, mentoring, teaching, coordinating, facilitating*, and *inspiring* or otherwise *motivating*.

A coach is someone who helps someone to improve in their work, operating both in an achievement-oriented and participatory style. A mentor is someone who helps another to improve their situation, which might include improving their work. That is supportive leadership. A teacher is someone who trains someone in a specific domain of knowledge. Good teachers not only train but operate as a coach as well, and sometimes a mentor, and so effective training often requires all four of the Path-Goal styles of leadership.

Of course, there is significant overlap in the roles of coach, mentor, and teacher. These are all forms of leadership, because the person being helped looks to the coach, mentor, or teacher for guidance, advice, knowledge, and insight.

Coordinating is a form of leadership in which one takes charge of the dependencies between what many others are doing and tries to optimally reorder them and inject steps to synthesize the various parts. A coordinator needs at least a little authority—either formal or informal—to get others to rearrange their work according to the coordinator's decisions.

A facilitator is someone who leads a group by coordinating the interactions among the group. Facilitators often lead discussions, during which they make sure that everyone gets a chance to be heard and that the discussion stays on point.

The use of facilitation is common in the Agile community. However, it is often assumed that the facilitator does not need to understand the topic being discussed. Yet, it is not clear how someone who does not understand a topic can determine whether the discussion is going off point.

Many in the Agile community also feel that if a facilitator makes a suggestion about the subject matter, they compromise their role as facilitator. But Socratic discussion uses hard questions, and posing such questions requires understanding of the subject.

One of us with expertise in a particular domain once worked for an organization facing serious challenges in that domain. Due to that organization's Agile practice's rigid definition of a neutral facilitator's stance, our expertise was ignored and minimized, to the detriment of the product under development.

If someone has expertise or good ideas, they should be free to share those!

Effective leaders often *do* manage to facilitate discussions well, even when they inject ideas of their own. What matters is how it is done, whether those in the room feel that the leader is open to challenge, and that the best idea will win, no matter who it comes from.

The need to inspire people is important to be able to truly tap people's full ability. Leaders who threaten people get only the minimum effort required to avoid punishment. If people's motivation derives from a belief in the inherent value of the work, because either it benefits them or it benefits others who they feel are worthy, then people will try their best. Motivation can also come from other needs, such as feeling of fulfillment or an opportunity to be creative. These are all positive sources, while fear of discipline is a negative one.

Applicability and Trade-Offs

A team of people is a group that is organized for a purpose. A team often needs many kinds of leadership. First, each person needs to be their own leader: each person has agency, and they need to take responsibility for their own outcomes, voicing problems, trying to get those problems solved to the best of their ability, and—if it comes to it—deciding whether to remain in the group. Some groups sometimes are not a good fit for a particular individual, and vice versa. That is just the reality.

Besides individual leadership over oneself, a team often needs leadership. For example, it might need coordinating leadership; it might need coaching leadership to help it to grow its abilities; it might need thought leadership about particular domains of knowledge that are mission critical; and it might even need some directive leadership for some issues.

Peter Drucker has said that an organization needs "an inside person, an outside person, and someone to get things done." In other words, one kind of leader is not enough. That does not mean that one person cannot fill all those roles; but such an individual is unusual.

By an "inside person," Drucker meant someone who can form relationships of trust with the members of the organization, someone they look up to. For a technology company, that individual might be a visionary CTO, or it might be a visionary product designer. In the case of Apple Computer, Steve Jobs was the CEO, but he was also the chief product designer and was widely seen as a visionary, so he was the inside person. When he returned to Apple, he was greeted as a savior, and many people worked at Apple only because it was led by Steve Jobs.

By an "outside person," Drucker meant someone to deal with the outside world. Any organization or team exists in a larger ecosystem, and that ecosystem can be leveraged, or it can undermine the organization or team. Managing the relationship with the world outside of the team or organization is essential. To do that, one must have some level of authority to be able to make promises, to negotiate deals, and to invest in resources. One must also be adept at managing expectations and coming across as personable, trustworthy, competent, and visionary to some degree.

The third role described by Drucker was "someone to get things done." That is the organizer. Within a team, an organizer might not need much or any authority, but the higher one goes in an organization, when a team member in turn oversees other subordinate teams, authority is increasingly important because without it, one cannot make decisions on behalf of one's subordinate teams. An organizer is someone who stays on top of everything, continuously watches for problems, takes action as soon as there is a problem, and orchestrates discussion and timely decision-making.

A recurring theme here is when to use one's authority and when to let subordinates decide. Hold that question in your mind as you read this, and we will address it at the end of this chapter, because it is the most important question about leadership, and it ties everything together.

Socratic Leadership

A Socratic leader casts themself as a fellow learner and asks questions and engages in discussion. The outcome of the Socratic process is getting the team to buy into the ideas they agree to as if they thought of them—and indeed they did, even if the leader felt certain about where the consensus would end up (or not).

A Socratic leader needs to understand the subject matter that is being discussed; otherwise, they are not able to ask deep questions, especially follow-up questions.

A particular challenge with the Socratic method is that some people are not good at thinking "on their feet," and so they might not be able to articulate their thoughts in real time. Thus, this method must not be used alone for discussing complex issues; those who need to think "offline" must be allowed to do so and then share their thoughts in a follow-up meeting.

Also, some people will not speak up in a group, and so a Socratic leader must watch for those who stay silent and proactively ask them for their opinion.

The Socratic method is time-consuming but leads to a deep understanding of the issue by all team members and a strong feeling of investment in the final consensus if everyone feels their contributions were fairly considered. Those who do not agree with the consensus will usually be willing to go along with it because the resolution was reached through a fair and logical process.

Sometimes a team is unable to reach a consensus. When that happens, if a resolution is needed, a leader might be required to arbitrate and make the final decision.

Voting is not recommended for important issues, because a vote indicates popularity of an idea, rather than determining whether the idea is best. If an arbitration is needed, the leader will need to decide and bear accountability for the outcome.

Servant Leadership

The term *servant leader* was coined by Robert Greenleaf in an essay in 1970.[16] He describes a servant leader as someone who becomes a leader by showing that they have the interests of the team at heart instead of their own and also by demonstrating competence.

Something that is often misunderstood about servant leadership is the idea that a servant leader is really just someone who facilitates or assists a team. In fact, Greenleaf clearly says in his essay that a servant leader *leads* and that the team *follows* but follows by choice.

A servant leader sometimes makes decisions that are not the choice of the team, but the team trusts the leader. In the book *The Servant*, James Hunter writes, "The leader should never settle for mediocrity or second best—people have a need to be pushed to be the best they can be. It may not be what they want, but the leader should always be more concerned with needs than with wants."[17]

There is a dilemma with servant leadership, in that the interests of the team might not be aligned with the interests of the organization; and so if the leader's authority rests upon the support of the team, then the leader might not be able to make choices that favor the organization at the expense of the team—at least, not too often.

An example is when it is time for pay raises to be considered. If one's leadership is based on popularity, then one will be incentivized to give everyone the maximum possible pay increase to maintain their support. Servant leadership therefore has an implicit assumption that the team is composed of rational, fair-minded people who are able to see and appreciate the big picture of the organization as a whole.

Theory X, Theory Y, and Mission Command

Management theory has a long history. The 1900s saw the emergence of scientific management through the writing of Frederick Taylor, Henri Fayol, and Max Weber. The 1930s saw the human relations wave through the pen of Elton Mayo, Fritz Rothlisberger, and W. J. Dickson. The 1950s produced "behavioral science," with authors such as Herbert Simon, Douglas McGregor, and Rensis Likert. So-called systems thinking included writing by Ludwig von Bertalanffy, Kenneth Boudling, Jacob Getzels, and Egon Guba. The 1970s saw authors such as James March, Karl Weick, and Johan Olsen.

Scientific management treated people as machines: so-called Taylorism sought to optimize each step of a workflow, so that each person was performing as much as they could, and thereby optimize the entire flow. It was presumed that a global optimum could be reached by optimizing each step of each task and assumed that all the tasks were predefined by a "scientific" designer of an optimal work process.

Taylorism did not actually seek to dehumanize people. In fact, Taylor believed that as each worker became expert in a task, they would be respected as an expert. However, that element of his theory is often forgotten.

Also, Taylorism did not consider the mental health of the workers or their personal motivations; it saw them only as mercenaries who needed to be pushed as far as they could bear, not too unlike the slaves of a galley ship, except that the shackle was replaced by the desperate need of a job in those times.

Max Weber wrote extensively about the need for a bureaucracy to bring order to management. Ironically, he viewed a hierarchical structure with rules of behavior as a solution to the favoritism and subjective judgment that was common within less formally structured organizations. Those times were characterized by what we, today, would call more Agile arrangements, and hierarchy and bureaucracy were seen as a remedy for the unfairness that was common in those setups.[18]

Systems theory as first described by Ludwig von Bertalanffy, Niklas Luhmann, and others pertaining to social systems viewed an organization much like a biological organism. In a systems view, an organization seeks a steady state, which authors of the time referred to as *homeostasis*, and there are feedback loops that maintain that state.[19] Understanding the organization means identifying and understanding the feedback loops. (It should be noted that today systems are viewed in a broader way, not necessarily requiring a steady state.)

Thus in original systems theory we had the beginnings of the notion that is so prevalent yet problematic today, that an organization exists in a steady state and that a "project" must be conceived to change the organization to a new (steady) state, with the return on investment of that change proven up front. Today's reality is that most organizations cannot be seen as being in a steady state, because the world around them is so rapidly changing.

Behaviorists such as McGregor and Likert rejected the idea that management was all about authority and control. McGregor labeled control-oriented management as Theory X and defined an alternative form, Theory Y, in which people prefer to act responsibly and so do not need to be tightly controlled, and they often apply creativity in their work, which benefits from looser control. McGregor and Likert's writing on this emerged during the 1950s—well in advance of the Agile movement.

Some in the Agile community think that before Agile, all organizations were run in an autocratic manner. That is not true. In fact, the debate over whether autocratic or empowering methods work best is very old. An empowering style of leadership that is still widely recognized as a model today is known as *mission command* leadership, which dates to the early 1800s and the Prussian army.

In the mission command model, decision-making is pushed down to the lowest level so that those in the field can act immediately, using their own judgment, without having to wait for orders. Mission command relies on training those in the field so that they have sufficient knowledge and leadership skills so that they can make competent decisions autonomously. The ideal field commander is an experienced leader and is knowledgeable of the "big picture" so that field decisions can take the overall strategy objective into account. Individual soldiers are expected to be able to make judgments too, so that if they are separated from their commander, they can act without delay.

Applying this in a business context, the mission command model informs team members of the big picture and goals and empowers team members to decide how to get the job done. Thus the Agile Manifesto principle that reads, "Build projects around motivated individuals. Give them the environment and support they need, and trust them to get the job done," is well aligned with mission command, from the early 19th century.

Unfortunately, this line of the Agile Manifesto has been interpreted by many as "always trust the team," failing to understand that the capability of the team members is highly relevant to how much autonomy a team should be given. Is complete autonomy for every team from day one what the authors of the Agile Manifesto intended? We do not know, but to us that would be irrational; and in the mission command model, a mission commander first assesses if the troops are up to the task or if they need more training. A mission commander also knows that there is a cost to failure—perhaps a cost in lives—and so failure as a learning tool best occurs during training, not during actual missions.

Good judgment about when failure as a learning tool is appropriate is essential, even for product development. If one's product is a website about retail items, some level of mistakes in the deployed system are probably tolerable, and so the risk of allowing people to "learn through failure" by working on actual systems might balance out as a valid instruction strategy; but if the product is the flight control system of an airplane or if it is a microservice that manages the bank account of customers, then the balance of risk and learning reward is different, and experimentation and learning need to occur far earlier and in a safer setting.

Knowing When to Intervene

If we could cite the one idea that screams most loudly in the Agile Manifesto, it is an idea that is not even stated explicitly but rather is embedded in the wording of each value. Each value statement specifies two things, and each statement is of the form "thing 1 over thing 2."

Then following the values statements, the manifesto says, "That is, while there is value in the items on the right, we value the items on the left more."

In other words, the four value statements emphasize balance and judgment. Nowhere does it say, "Do this instead of that." In every statement, it says, in effect, "Use your judgment about the situation."

This has been overlooked in the community, where the polarity between the two statements is taken to mean that one side is bad and the other side is good. The caveat of the need for judgment is so frequently overlooked that many written interpretations of the Agile Manifesto leave it out entirely and pose a series of good/bad opposites.

We mentioned earlier that since the Agile movement began with Extreme Programming (XP), which set the tone for extremes, this could be why the Agile Manifesto came to be interpreted in an extreme way: when people read the value statements, they read them as, "Do this instead of that," rather than "Consider these two things and balance them for the situation." Agile became a movement of extremes, to its detriment.

Applying contextual judgment was actually a disruptive idea at that time, and so extremes were not needed to displace the unhealthy practices that had become common in the industry. Practices such as creating a huge and inflexible task-level plan up front and defining all of the system's requirements during a requirements phase and then passing that off to a set of "implementation teams" were extremes, and so bringing judgment back was the most novel idea.

Just as the right kind of leadership is the most key thing to have, judgment is the next most key thing. Judgment follows rather than precedes good leadership because a good leader must exercise judgment. A good leader uses judgment when deciding whether to let their teams decide something on their own or when intervention is necessary.

Good judgment cannot be taught. It is innate in some people, but experience and knowledge improve everyone's judgment. That is why any form of leadership that needs to apply judgment is more than just facilitation. People need to decide how best to do their work—most of the time, but not all of the time. Knowing when "how" matters is a matter of judgment; and deciding whether the situation can tolerate learning through failure and when it cannot is also a matter of judgment. Deciding what kind of instructional mode to apply—coaching or teaching—is also a matter of judgment.

Judgment occurs at every point. Judgment happens every day. Judgment is what it is all about. Good judgment leads to good outcomes.

We now have a foundation for understanding leadership in its various forms. We will use that foundation in the rest of this book for considering what forms of leadership are needed in various contexts, with agility in mind.

We hope that you now see that leadership is a complex and multifaceted issue. A team needs leadership—not simply one kind, but many kinds. Leadership is not only needed for a team; it is also needed for teams of teams, at every level of the organization.

Notes

1. www.youtube.com/watch?v=Opnk-cPOM50&app=desktop
2. Mark Schwartz. *A Seat at the Table*. IT Revolution Press. Kindle Edition.
3. Titus Winters; Tom Manshreck; Hyrum Wright. *Software Engineering at Google*. O'Reilly Media. Kindle Edition.
4. www.teslarati.com/elon-musk-method-tesla-explained
5. www.wsj.com/articles/quibi-was-supposed-to-revolutionize-hollywood-heres-why-it-failed-11604343850?mod=mhp
6. en.wikipedia.org/wiki/Path%E2%80%93goal_theory
7. fortune.com/2016/03/04/management-changes-at-medium/
8. www.wired.com/story/silicon-valley-tyranny-of-structurelessness/
9. www.fastcompany.com/3022131/does-google-need-managers
10. en.wikipedia.org/wiki/Iron_law_of_oligarchy
11. www.bbc.com/worklife/article/20200827-why-in-person-leaders-may-not-be-the-best-virtual-ones
12. Marta Wilson. *Leaders in Motion: Winning the Race for Organizational Health, Wealth, and Creative Power*. Greenleaf Book Group Press. Kindle Edition, p. 134.
13. pubmed.ncbi.nlm.nih.gov/27194144/
14. hbr.org/2018/06/too-much-team-harmony-can-kill-creativity
15. en.wikipedia.org/wiki/Leader%E2%80%93member_exchange_theory
16. www.goodreads.com/book/show/12552292-the-servant-as-leader
17. James Hunter. *The Servant: A Simple Story About the True Essence of Leadership*. Crown Business, 1998, page 66.

18. From Theories of Modern Management, by W. Hal Knight, chapter 3 of *Principles of School Business Management*, p. 54. Available online at files. eric.ed.gov/fulltext/ED282284.pdf

19. Bruce D. Friedman; Karen Neuman Allen. *Systems Theory*. SAGE Publications. Available online: www.sagepub.com/sites/default/ files/upm-binaries/32947_Chapter1.pdf

4

Ingredients That Are Needed

As explained in the preface, the purpose of this book is not to serve as a textbook. We therefore will not provide an exhaustive explanation of all of Agile 2's principles.

Instead, now that we have explained why Agile 2 is needed and laid out some of the problems with how Agile is defined, interpreted, and practiced, we can start to assemble some core ideas for what Agile should look like—in other words, the insights that underly Agile 2.

We also want to stress that there were some good ideas in what we might call "traditional Agile." We do not want to lose those. But as explained, many of those ideas have already been lost through extreme—and we believe incorrect—interpretations of the Agile Manifesto or through corruption of the ideas by agendas and special interests within the Agile community.

Elements to Keep or Adjust

Most people do not know that the portion of the Agile Manifesto that is copyrighted is the "Values" section. The "Principles" section, which is on a second page, is not. We have been told by multiple sources that the values were created during the meeting at Snowbird. The principles were created through email discussions that occurred during the following weeks, and it is not clear that all of the 17 Snowbird participants actually weighed in on those principles.

Even so, most people in the Agile community treat the principles and values as having equal authority, and so we will too.

In addition to being defined by the Agile Manifesto, traditional Agile is defined by the practices and ideas that have come to be commonplace within the Agile community at large. These include the Extreme

Programming, Scrum, and Kanban frameworks and their practices. Additional ideas have arisen over the years and are seen by most people as associated with Agile, although they are usually treated as adjunct and not core to Agile.

There is controversy and disagreement about the general applicability or effectiveness of many of these adjunct ideas, while in contrast most people in the Agile community strongly buy into the core ideas and practices (many of which we challenge in this book). Some of the adjunct ideas are mob programming, Open Space meetings, Lean coffees, and many other approaches. Agile scaling frameworks such as Scaled Agile, LeSS, Enterprise Scrum, and others are usually considered to be additions to Agile that some Agilists support and others do not.

The Agile community has also adopted ideas and practices from other areas of thought, most importantly Lean thinking, systems thinking, the Tuckman team model, Virginia Satir's version of the McKinsey J curve (usually referred to as the *Satir change curve*), and DevOps. Organization models that operate mostly through self-organization, such as the holacracy model, are also often seen as a tangential aspect of Agile.

We are going to focus on the core elements of Agile. Whether they are stated in the Agile Manifesto or are considered core because of their widespread acceptance does not matter to us; they are part of the reality of what is Agile today. Many of these have great value, although in some cases either they are not well understood and thus are misapplied by many practitioners or they are actually flawed to some degree but can be saved by adjusting them.

Collaboration vs. Handoffs

One of the great successes of Agile is that it discredited the idea of the *handoff* as a collaboration practice. A handoff is when one specialist completes their task and hands the piece of work to another specialist.

Handoffs are not necessarily bad. In an automobile assembly line, work progresses through a series of handoffs. The Kanban methodology operates through handoffs: as each step of a Kanban flow completes, the piece of work is handed off to the queue for the next step. The powerful concept of a pipeline, which is a central feature of DevOps, is based on handoffs between steps.

The reason that handoffs *can be* suboptimal are many.

- **Blame**—If each step—each expert—is measured independent of the end-to-end outcome, then each step competes with the other steps. If something goes wrong, downstream steps have an incentive to blame upstream steps. Success becomes a zero-sum game of each step trying to prove that it was done right but the other steps were not. Collaboration and trust between steps break down.

- **Local optima**—Each step tends to learn little about the other steps, so each step tends to optimize its work for its own purposes instead of for the best end-to-end outcome.

- **Loss of continuity across steps**—When a step completes, those involved tend to quickly forget about the piece of work that they completed so that they can focus on the next piece of work. As a result, when downstream steps need to discuss the piece of work, they need to interrupt an upstream step, which is no longer interested or focused on that piece of work. In fact, the individuals who worked on it might have been shifted to another job function; and if they were a contractor, their assignment might have ended and so they have left.

- **Bottlenecks**—If one step in the sequence is slower than the others, then the entire flow will be limited. This can be addressed by giving each step a sufficient number of people so that each step has a manageable level of "work in progress" or work waiting for the next step.

- **Rework is out of context**—If a downstream step discovers a problem and the work item has to be returned to an earlier step, it disrupts the flow, and the person who must make the repair might have a poor recollection of the work item, since they worked on it some time before, perhaps days or weeks before.

- **Delay**—Between each step there is a queue where work waits until the next team can do it.

- **Handover costs**—If each work item is unique, then transferring work from one person to another can incur a cost because the second person must learn about the unique aspects of the work item from scratch.

On the other hand, a handoff process has benefits:

- **Flow**—Coordination of work steps is naturally achieved, creating a continuous flow.

- **Balance**—Work can be balanced by increasing the throughput of slow steps, so that the flow is continuous.
- **Parallelism**—A handoff can branch into parallel flows, and reconverge later if necessary, achieving parallelism.

Deciding whether a handoff is the best way to organize work depends on the circumstances. An assembly line works well for Toyota, which uses the approach of a car moving from station to station with only a few steps performed at each station—usually a single person.[1] Ferraris are not made quite that way: each car progresses through a series of production phases, and each phase is performed by a team, with the car stationary while that team performs many more steps than are performed in a Toyota station.[2] What works well for one kind of car is not optimal for another kind.

One thing is clear: even when a handoff process is used, it is helpful if the people in each work step take an interest in the other steps. That way, they can partially avoid the "local optima" problem. Also, if someone performing one step suggests a change to the flow, the discussion can be more informed, because everyone has some level of understanding of how each step is done. This might be an argument for rotating people among the various steps, which might be feasible in some situations but not others. For example, if one step is the creation of a custom neural network model, it could be that only a subset of the team members have the expertise needed to make model changes.

The value of taking an interest in other steps is huge. It has long been common for business-focused stakeholders to say, "That issue is technical. I don't need to know how it will be done." That argument often does not hold anymore. Today, the "how" of our work is often an important part of the value proposition, and there are often trade-offs between having additional business features and making improvements to how something is done, and so all stakeholders need to understand all the factors involved in a trade-off between the new business features and the proposed improvements to how the work is done.

In a LinkedIn discussion thread, a manager and DevSecOps lead at a large technology company wrote this:

"Stakeholders should be able to assume that the organization building the product knows how to properly build the product, and also that the organization doing so will properly inform the customer of the customer's options."

This is a valid perspective if the way that the product is delivered does not have market value. For example, in government contracting, where a proposal is accepted and there is a well-defined schedule, then the how does not matter. The customer will be satisfied if the work is completed on schedule and things work.

However, this is not how many industries work today. Consider, for example, the commercial rocket launch industry. In that industry, it is rare that anyone pays a company to design and build a rocket. The rocket designs exist, and customers merely sign contracts with launch providers to put a payload into orbit (or beyond). The providers that can demonstrate rapid turnaround, reliability, and low cost—and that can boost the size payload required—are most favorable. SpaceX has shown that it can outpace other companies, because the way that it builds its rockets is so superior. It uses a highly iterative and fast turnaround approach, and it also designs the rockets to operate as appliances so that launches are simple and SpaceX can perform many launches in rapid succession without overtaxing ground crews.

In short, the way that SpaceX works is strategically important and has customer value. SpaceX's president Gwynn Shotwell knows this well when she markets SpaceX's products and services to customers. She can say things like, "We have rapid turnaround, we are continuously refining our designs through frequent improvements, and so our products just get better and better at a rapid pace."

Here is what she actually said recently: "We are focused on rapid innovation and relentless focus on progress . . . . In 2017 we broke the record for launches. We launched last year more than any country or company, with 18 launches. We should beat that record this year."

Our point is that Shotwell, as president and head of sales, not only understands what their rockets do, but she understands how they are built, and she brags about it and touts their methods as a major source of value for customers and a reason why SpaceX's products are always getting better and why SpaceX should be trusted in a high-risk industry.

Empowerment

The Agile community believes strongly in the empowerment of teams to make their own decisions. This is a good thing. In a previous chapter, we explained some leadership models and in particular talked about Theory Y management and the mission command leadership approach. These approaches also emphasize the autonomy of people in the field, rather than relying on central management for decisions.

Agile 2 Principle: Technical agility and business agility are inseparable: one cannot understand one without also understanding the other.

Agile 2 Principle: Business leaders must understand *how* products and services are built and delivered.

There is a gap, though. The Agile community often treats autonomy as a given—that from the start, a team should be seen as autonomous. In contrast, mission command requires a leader to assess a team's ability and then give the team autonomy in the areas in which they have shown competence. Assessment is an important part that is often forgotten (or never learned) by Agile practitioners.

For example, if a team is composed of highly experienced members and has shown that it can deliver product features in a way that meets all of the organization's criteria, then it can probably be trusted to work on its own, autonomously. In contrast, if a team consists of newly hired college graduates, none of whom has ever worked on a product in a professional capacity, then considerable oversight is clearly needed. That does not mean the best approach is to tell them what to do and micromanage them, but it would be foolish to just assume that the team can operate on its own.

There is a related concept of trust. Many Agile practitioners say that one should "trust the team." We have heard this stated as an axiom and an absolute; equivalently, the "team always knows best." This is also contrary to the mission command theory of leadership. Autonomy is given when a team is trusted, and trust is contingent on assessment of capability.

Thus, trust is not a given but must be justified. As Johanna Rothman has said, "When your management asks you for an estimate that is more than an order of magnitude estimate, it's because they don't trust you. That's because you haven't been delivering often enough for them. . . . You build trust by delivering. Period."[3]

What she means is that people continue to ask for details if you have not yet earned their trust. Once you earn their trust, they are more likely to accept your estimates without details, because you have created a history of delivering what they need while not overrunning cost constraints.

When people have autonomy in how they accomplish given objectives, they feel empowered and inspired. They feel they have control over their success, which gives them hope. They feel trusted and therefore valued. However, autonomy should only extend to areas for which a team is ready. Giving a team too much autonomy might be setting them up for failure. If that is by design, as a learning exercise, then the goal is the team's learning; but if it is essential that the team not fail, then they should not be set up for failure, and consequently they should not be given more autonomy than they can handle.

Teams also sometimes do what is in their best interests, rather than the organization's. Consider programming languages. If a team is given a choice of which programming language and development frameworks to use for a product, will they make choices that maximize and balance reliability, maintainability, robustness, performance, and productivity appropriately? Or will they make choices that maximize their personal marketability, as well as their personal productivity, at the expense of the other attributes? Be your own judge, but remember that today product developers in the US Silicon Valley region tend to change jobs every few years.[4]

Also, consider that younger programmers tend to favor "dynamic" languages that make their jobs easier, whereas older and more experienced ones tend to prefer

Agile 2 Principle: Favor *mostly* autonomous end-to-end delivery streams whose teams have authority to act.

"static" languages that are more maintainable and more robust. Python is extremely popular among young developers,[5] even though it has been shown to be difficult to maintain.[6] Python is a language that enables one to code quickly, because it requires so little in the way of defining one's intent—something that one pays for later when others need to make changes to the code. The skewed preference among younger developers for "make my job easy" languages argues strongly for some oversight by senior developers who have a lot of experience and who can be trusted to balance issues such as productivity and maintainability. Those senior developers can be team members, however—they need not be overseers from the outside.

There is also the matter of team versus individual. The Agile community has come to speak of "the team" as the smallest unit of activity within an Agile initiative. However, the smallest unit of activity is actually the individual—a person. People on a team do not all operate in synchronization. The ideas pertaining to autonomy, trust, and experimentation apply equally to individuals as well as teams. Someone on a team might be ready to write a certain kind of code, but someone else might not be. If someone is not, either they should not be allowed to write that kind of code or they should be paired with someone who can supervise.

Not everyone can learn everything, however. Learning something sometimes requires a foundation that some people have and others do not. We discuss this important factor when we talk about experts and generalists.

> **Agile 2 Principle:**
> Individuals matter just as the team matters.

Experimentation

Another idea that is connected to common lines of Agile thought pertains to experimentation and the idea of "accepting failure." However, failure always has a cost, and it is a leader's responsibility to assess if the cost of failure outweighs the learning benefits of allowing a team to fail and thereby learn from their mistake. Allowing failure should not be an absolute but is a matter requiring good judgment. Experienced Agile practitioners know this, but sometimes we hear an absolutist interpretation from those less experienced that any failure is learning and should be accepted.

It should be pointed out that it is not only teams that fail; it is just as often leaders who make decisions that lead to failure. However, a leader also belongs to a team, such as a management team, and that team's leader needs to view failure in the way that we are describing here.

If a leader of a team makes a bad decision, saying that the team would have known better if the leader had stayed out of it is not the right way to look at this situation, because it is just as likely that the team would have made a bad choice. As we have said, there is no fool-proof way to ensure that the people who would make the right choice are the ones who get to make the choice. The fact is, mistakes happen and will always happen. How we handle them is just as important as trying to prevent them.

Tolerating failure really means that one should expect—and accept—that if an organization is moving quickly, there will be mistakes. Unlike the old days when organizations planned carefully for every change, today's setting requires rapid change and rapid pivoting, which means that it is impractical to plan everything in detail. Often it is better to just plan a little and then try it as a contained experiment and see what happens. That enables the product teams to learn about the market and the technology, which will lead to fewer mistakes and also help to prevent large and costly mistakes.

SpaceX does this for its rockets. One would think in a high-risk enterprise such as building rockets that any company that does that would spend years perfecting a design, simulating it, and then carefully crafting

Agile 2 Principle: Product development is mostly a learning journey—not merely an "implementation."

it. That is not actually how it is done, however, and SpaceX is not the first to use Agile methods for rockets.

The aerospace industry has a long and rich history of creating early proof-of-concept designs, trying them, measuring, and then "going back to the drawing board." The Apollo rocket's F-1 engine is an example: engineers spent years trying one modification after another to get the engine to be stable. The problem was that the size of the engine created instabilities in the fuel flow into the combustion chamber, creating vortices that would spin out of control in a microsecond and cause an explosion that destroyed the engine. In those days, detailed computer simulation was difficult to do, so trial and error was the only approach available. Finally, the designers were able to come up with a set of baffles that stabilized the flow, and the engine then took astronauts to the moon.

At SpaceX the engineers take this iterative approach much further. They use a "hardware-rich" approach, whereby they make a design change to a part, inspect the entire component assembly in real time using advanced 3D modeling software, and then click "print" to make the part and then try it in a real engine. As a result of this process and the ability to rapidly "print" metal parts, the turnaround time—their design change cycle time—is days instead of weeks. This hardware-rich approach to design and construction is what enables SpaceX to create version after version of engines and entire vehicles, rapidly experimenting and improving as they go. In the course of that, many design changes will fail—they will prove to be unworkable—but that is not seen as a bad thing; it is seen as part of the refinement of the design. No one is blamed; it is viewed as normal.

Accepting failure in the course of experimentation and iteration is essential. It is not the same thing, however, as blanket acceptance of failure. Failure always has a cost, and if someone has a history of expensive failure after expensive failure, it might indicate either bad luck or recklessness: perhaps things are not being thought through, and there has been too much rush to push something to the market in a way that was not contained enough.

That is a judgment that leadership must make. Failure is an important element of learning and improving products, but it is not a license to be careless.

Feedback Loops

We learned that systems theory emphasized the identification of feedback loops for understanding organizations. Organizations are viewed as being in a steady state, and it is feedback loops—intentional or unintentional—that keep them there. We also learned that today's organizations are rarely in a steady state, but the concept of feedback loops is still an important one.

The Agile community frequently speaks of feedback loops, often referring to a feeback-based approach as *empiricism*. The Scrum methodology defines a feedback loop from the customer to the Scrum Product Owner, who is the person charged with defining what the product should do. There is also a feedback loop in the form of a retrospective process that a team undergoes at the end of each product development iteration (usually a two-week period), which Scrum refers to as a *sprint*. In a retrospective, the team reflects on how well they are working and if they should adjust any of their practices.

The DevOps movement arose either as part of Agile or in parallel with it, depending on who you ask. Either way, DevOps generalizes the feedback loop concept and emphasizes that feedback loops should be used everywhere to ensure that a product development process flow is self-correcting—that is, so that it is always checking itself to make sure that quality is maintained, no matter how fast the process runs. This is why DevOps is fundamentally a risk management theory: it replaces the use of manual checkpoints for controlling risk with a continuous process that manages risk automatically. We will discuss it more from that perspective in Chapter 9.

DevOps feedback loops can take many forms. These include Agile feedback loops, but they also include the use of metrics on product usage to inform on how features are actually being used, metrics that inform how well the product is performing, metrics that show in real time the cycle time and lead time between various steps in a product development and test flow, and metrics that indicate how often various quality practices are being used.

An important dimension of DevOps is its focus on the product rather than on the team. DevOps agrees with Agile that ideal teams are autonomous and self-orga-

Agile 2 Principle: From time to time, reflect, and then enact change.

nizing, but organizations that practice DevOps well do not have fully autonomous teams and do not rely fully on self-organization. They have specialist teams who assist development teams, and they have leaders who make sure that groups of teams are coordinating. They also have invested heavily in self-service tooling to enable teams to be more autonomous.

For example, Google developed an in-house tool that automatically identifies all of the products that interact with a given product, and when a

change to that product is committed, the system first automatically loads the automated integration tests of all of the dependent products and verifies that all the tests pass. That level of automation, which checks on interactions between products maintained by different teams, enables teams to be more autonomous. This kind of automation is a sophisticated feedback loop, and it is one that most organizations (other than the likes of Google) cannot afford to create, so it is not unreasonable to assume that the level of autonomy of teams at Google is probably higher than at other places.

What works for one organization might not work for another.

Incremental Delivery Is Powerful

Extreme Programming encouraged continuous development and internal release of software to prove that the code is always working and therefore prove continuous progress. The Agile Manifesto encoded this idea in its principle that reads, "Deliver working software frequently, from a couple of weeks to a couple of months, with a preference to the shorter timescale."

This is a powerful idea, and it got us all away from the practice that had become common by the late 1990s, in which projects were divided up into phases and no code was expected to work until the end; what often happened was that everything was "green"—in other words, all status reports showed that things were on track—until the day of final delivery, when it was revealed that nothing worked and six more months were needed. Some people call this a *watermelon* project—green on the outside and red on the inside!

By demonstrating a working system all along, each time with more features complete, one can actually see progress. This approach makes product development more like building construction, where you can actually *see* progress with your eyes.

The problem with the Agile view of cadence-based delivery is that it is phrased in terms of working software. Working product (software or otherwise) is important, but what is even more important is business outcomes, when new product versions are deployed for use. We will talk about that later in the section "Outcomes Matter More Than Outputs."

A Sustainable Pace Needs to Include Uncommitted Time

Agile teams typically work against a backlog of features to build, drawing from the backlog to obtain work requests. This means there is always more work to do. The work is never done, because such backlogs always grow. None of us can recall ever seeing an Agile backlog become empty. For those who use Scrum, the Scrum Guide even says that the backlog will never be empty for the life of the product. An organization might decide to not continue work and shift the teams to another product's backlog, but this is almost never because the backlog became empty.

Contrast this with the flow of work in a waterfall project. We use waterfall as an example to illustrate a point, not to advocate for waterfall, which is best suited to repeatable work that can be fully defined up front. For our example, suppose that the project is to build a new product—something that we would not typically suggest waterfall for.

A waterfall project has phases: there is a requirements phase, a design phase, an implementation phase, and so on. A phase is marked by completion of all the deliverables defined for that phase. For the early phases, the deliverables are documents. When a phase begins, there is very little pressure. This is because there is no deadline looming, and one's main task is to read and mentally digest the deliverables from the prior phase and begin thinking about how to approach the current phase.

The pressure picks up later, as the phase's deadline approaches. In fact, there can be late nights and a lot of stress to meet a deadline. Thus, waterfall phases are characterized by beginning with little or no pressure, but the pressure ramps up continuously until the deliverables are finally actually delivered. Then the pressure vanishes, and one waits for the next phase to begin. There is often a gap of days or weeks between phases, while stakeholders review deliverables.

As you can see, waterfall projects have periods of low pressure and periods of high pressure. In contrast, an Agile initiative has constant pressure that never lets up. A common pattern for Scrum teams is that the pressure begins at a low level when a development sprint starts and then escalates toward the end of the sprint. Then the team plans and starts again. By comparison, teams that use Kanban have constant pressure, although the pressure is maintained at a low level, but the pressure never goes away. Scrum teams are under constant pressure too, even though it waxes and wanes a little through the course of each sprint.

Constant pressure is not healthy for many reasons. People need to have pressure removed for periods of time. A weekend off does not accomplish that, because there is the ever-present worry in the back of one's mind about the state of one's work. Holidays help, but they are not sufficient if one is under constant pressure for the months in between holidays.

This is why Agile work should be organized to purposefully inject breaks from pressure. There are many ways to do that, such as by dedicating periods during which team members are asked to "tidy up" their work but with no actual deliverables or by giving them periods during which they are asked to be creative and learn new tools or work on proof of concepts of their own.

Business stakeholders might see this time as unproductive, but that is short-sighted, because these breaks help to keep team members fully engaged mentally in the long run and enthusiastic about their work, instead of becoming burned out.

Elements to Add, Add Back, or Change

The Agile Manifesto and the Agile community have missed some important things. During the early part of the 2010 decade, the community had realized that methodologies such as Scrum could not work in a large organization unless the organization changed to accommodate it. As a result, discussion within the community about organization change ramped up. The mantra "Don't *do* Agile, *be* Agile" arose. The idea was that an organization had to become Agile in order to enable Agile teams (usually Scrum teams) to work. Yet, by and large, the community had few answers for what it meant for an organization to "be Agile."

By 2014 some of us noticed this confusion, and we formed a group of organization change management experts and Agile experts and collaborated on what it meant for an organization to "be Agile" and how an organization could get there. We published our ideas on the website Transition2Agile.com.

Meanwhile, the DevOps movement was growing, and the Agile community was starting to discover ideas from other disciplines and enrich thinking about the organization scale questions. So-called scaling frameworks started to receive a lot of attention, and these divided the community, with some people embracing certain frameworks, and other people embracing others. The landscape of ideas about how to scale Agile became a set of commercial territories.

We are not going to wade into all that. Instead, we are going to list some important ideas that we think need to be present for Agile to work—ideas that are missing from the core ideas of Agile and are not even present among the adjunct ideas that are common. This does not mean that these important ideas are never voiced. In fact, they often are, but they are most often voiced in a dissenting way, by authors who are pointing out gaps or problems with Agile.

Leadership Is Essential and Is a Multifaceted Issue

The Agile community embraces the idea that everyone is a leader. Agile advocates for self-organizing teams, which implies personal agency and emergent leadership. What Agile fails to acknowledge is that leadership is a complex matter, and there are many forms of leadership—many more than those that are hinted at by the Agile Manifesto or described by the Scrum Guide, the two sources that are most often pointed to for guidance on Agile leadership. (Fortunately, more thoughtful discussions of leadership have been happening recently in some segments of the Agile and DevOps communities.)

We have already explained why autonomy is not an absolute: it must be given based on one's assessed abilities and also on the degree to which learning through trial and error can be accommodated in the given circumstances.

Self-organization implies autonomy; but in earlier chapters we explained that when a group of people self-organize, leaders often emerge, and informal authority develops. So when a team self-organizes, it is common that it ends up with de facto leaders, either good or bad. Therefore, even when allowing teams to self-organize, it is essential to understand leadership issues, monitor things, and be ready to intervene if necessary.

The Tuckman model, which is often cited by the Agile community as a model for how teams self-organize, clearly states that teams can easily get stuck in a "storming" state and require intervention to enable them to progress to eventually become "performing." Thus, such teams are not actually fully self-organizing, but rather are mostly self-organizing with occasional intervention by someone with authority.

There is also the argument that intervention need not have authority, that it can be through a facilitator who helps the team to understand why it is stuck and enable the team to decide on action to move forward and collaborate better. But this is not guaranteed to work. It brings to mind a situation that one of us experienced during a divorce, when we were sitting with our spouse (soon to be our ex-spouse) and a court-appointed moderator who facilitated the discussion, but no agreement was reached, so the matter went to court.

Facilitation does not work when parties have conflicting values, agendas, or beliefs; in that case, someone needs to arbitrate, which means decide, which requires authority.

The biggest challenge with leadership is to be able to hold back but guide and catalyze. The most effective and lasting progress occurs when those doing the work devise the ideas that advance the work. If a leader proposes the ideas, those doing the work are less invested—it was not their idea. If someone is implementing their own idea, they are, in effect, proving their own idea and can expect to receive some congratulation and recognition for the success of their idea.

Product developers are keen on receiving credit for their ideas. Some of us have had experience in a sales or sales support capacity and noticed that salespeople often do not care whose idea something is. All they care about is getting credit if the sale is closed. In contrast, product developers are craftspeople and creatives, and they care deeply about their ideas and getting credit for their ideas. To not get credit would be like an artist not signing their painting.

That is why helping product developers to come up with an idea on their own is so powerful. They live for their ideas. If you help them to define the idea that leads to success, you have won their heart and their commitment to the work and your leadership.

So the key is to help people to find the ideas themselves. That is a subtle process. It often requires questions and some suggestions about a line of thought, leading to a promising path. If you suggest the basic path to an idea, but the developer still devises the actual idea, it is still *their* idea, and you have helped them to create an idea that they can own.

On the other hand, sometimes people are wrong. This is where tolerating failure comes into play. Can you afford to let them try their idea, even if you are sure that it will not pan out? Or is the cost too high? Or is there a chance that you as their leader—presumably with more experience—are wrong? That happens too. One of us has experienced many times when our boss—in one case a CTO and in another a development director—told us the path we were taking could not work and that we could not do it, but we proved them wrong.

One of the hardest decisions a leader must make is when to use their authority and override what a team member has decided. There is a cost to that: that team member might trust their leader and still feel committed, but then the idea is not theirs and so their commitment is inherently less. But if the leader's approach does not work out, whose fault is it? It is absolutely essential that if a leader tells team members to use an approach and that approach does not work, that the leader explicitly acknowledge the mistake and say clearly that they were wrong. Otherwise, the team will feel that they are taking the blame for the mistake, and they will then start to think of how they can protect themselves from the leader's decisions.

There are also areas in which a team might want their leader to make the decision. Generally these are things that affect the whole team but are not directly related to the work itself and perhaps affect how everyone works together. For example, people on a team often want someone to "get things organized." Not everyone on the team wants to participate in the setting up of things, although some might. Many people on a team are often anxious to get on with their work and do not want to be involved in setting up behavioral norms or processes. They care about those issues but might prefer that someone they trust takes care of setting that up. On the other hand, if things are set up poorly, they will want to be able to voice their concerns and be listened to.

Teams also usually want their leader to effectively represent and advocate for the team. Teams usually do not want to be distracted by the intricacies of how resources for the team are obtained, how agreements with other services are made, and so on. But again, if something is not right, they want to be able to voice their concerns and be heard. Therefore, a leader must talk to their teams and ask what is working and what is not, taking any concerns that are voiced seriously and continuing to update the teams on progress in addressing those concerns.

In a previous chapter, we explained various kinds of leadership, but in this chapter so far we have only discussed individual leadership and leadership with general authority over the team. Teams also usually need other kinds of leaders within the team. These other leaders may have explicit authority in some matters, or not; in either case, the scope of their leadership is narrow and pertains to certain kinds of issues.

Unless you live in an authoritarian regime, authority is never unlimited. Your work supervisor or team lead cannot force you to stay in a job if you do not like it, and they cannot tell you how to live when you are outside of work.

Even within the context of the job, authority can be very specific. A team lead is usually someone who has final say over team decisions in general, even if they act as a servant leader and rarely use that authority but instead act in a more facilitative or coaching manner. If a team has such a person, they are usually called the *team lead*.

Teams also often have leaders (appointed or emergent) who provide thought leadership or have responsibility for specific processes. For example, there might be a technical lead who is the arbitrator and possibly the visionary for how the technology will be used by the team or teams. There might also be someone who is given authority to perform actions such as deploy to production or access production systems to diagnose a problem. In those cases, the authority is narrowly defined to those areas—the technology or the deployment, respectively.

Leadership is a complicated issue, and leadership relies entirely on contextual judgment at every turn.

Leadership is also very personal: everyone has their own style of leadership, even if they use certain models as a guide. There is a great amount of nuance to leadership, much of it tacit. There is no repeatable pattern that everyone can simply execute. Leadership is complex because humans are complex. Trying to define the perfect model of leadership is like trying to define the perfect relationship—all we can say about that is "good luck!"

> **Agile 2 Principle:** The most impactful success factor is the leadership paradigm that the organization exhibits and incentivizes.

For leadership, at best we can define models and describe some of the considerations for those models and then say again, good luck!

Include the Customer

As we explained, Agile is a collection of things—Extreme Programming (which came first), the Agile Manifesto, Scrum, and many ideas and practices that have either emerged from or been adopted by the Agile community over the years. Not all of those things agree, which is normal for any idea space.

The Agile Manifesto clearly puts an emphasis on the customer through its third value, which states, "Customer collaboration over contract negotiation," and through its first principle, which reads, "Our highest priority is to satisfy the customer through early and continuous delivery of valuable software."

Yet somehow the customer has become secondary in the way that Agile is practiced by many organizations. David Cancel, CEO and co-founder of Drift, said in a talk, "I hate Agile, and I hate it because. . .it's just another abstraction layer. We start to talk in weird made-up language like points and velocity, and never do we say the word *customer*. Never is customer involved in the process," and a room of 1,400 people applauded.[7]

The Scrum methodology defines the role of Product Owner (PO), who is the chief business representative who deals with a Scrum team. The underlying assumption is that the PO wants the system, and so presumably they know what it must do. They are presumed to have a "vision" for the software and are the one who is tasked with defining the requirements, usually in the form of *user stories*, which are short outcome-oriented feature descriptions.

The problem is that the PO is often not the actual customer or user of the system being built. Some Scrum advocates will say that is "badly done Scrum," and they are right, yet it is a common situation. That is bad enough, but in many large Agile programs the PO is often not even a business stakeholder but is actually an IT analyst or manager who liaises with business stakeholders. As a result, the customer is often not a participant in the process. In addition, the customer is often not an actual user of the system. The customer might be someone who wants the system to be built so that a different group of people can then use it.

Salespeople understand these different kinds of stakeholder: they differentiate between the customer or decider, the buyer, the user, the blocker, and the advocate or champion.[8,9]

So, what is really needed is not *customer* participation, but *user* participation, so that the team(s) can obtain early and continual feedback from actual users of the application that is being built.

Some Agile teams remedy the absence of the user by explicitly including real users as trial users at various points during development. That is what is needed: real user participation throughout development, whereby users try new features as they are completed, instead of waiting until some point in the future at which an entire release has been completed and it is too late to modify features being developed or reprioritize them.

The customer—the person paying for the system—is another matter. If the PO is not the actual customer, the risk is that the PO will not have the actual vision that the customer has.

Agile 2 Principle:
Obtain feedback from the market and stakeholders continuously.

The PO role presents one of the scaling problems with Agile methodologies as originally defined. A PO is supposed to work directly with each team. But if a product requires many teams, one PO will not have time to spend with each team, and so one will have to use proxies for the PO—often business analysts. There is nothing wrong with that if the PO stays on top of things and personally views the features as they are being developed to make sure that their proxies are accurately representing the vision. This can work fine as long as the PO is actually the owner of the vision, and if real users are giving feedback.

Too often the PO is not the owner of the vision; the PO is often someone who has been handed a directive to create a particular kind of product that has been defined by a marketing team. That treats the PO as an order taker rather than a team member because the PO has not been involved in the process of defining the vision. The PO then lacks all the thought process that went into the product vision, including market-facing experiments and falsified hypotheses. The PO is likely to be lacking the internalized tacit knowledge of what the product needs to do to solve the user's problem—to help the user to do the "job to be done." The PO is then just a manager, not a visionary, but the Scrum process requires the PO to be a visionary, because the PO must define the requirements and weigh trade-offs about features, guided by the vision.

A common situation is that the PO is appointed by and reports to a project manager. When that is the case, the PO will be inclined to behave as if the project manager is the customer because they are managing the PO. As a result, the PO does not stand up for the needs of users when it conflicts with the up-front plan developed by the project manager.

The Agile Manifesto states the ongoing involvement of the business stakeholders—the visionaries—more flexibly. It says, "Businesspeople and developers must work together daily throughout the project." This allows the possibility that the visionary might not be able to work with each team but might delegate others. The manifesto leaves open how this is achieved.

The following are essential:

- Actual users need to participate throughout product development; otherwise, teams are not benefiting from early feedback about features by those who will actually use those features.
- The primary business stakeholder needs to participate in the development of the vision for the product and direct the product priorities on an ongoing basis.
- Individuals from the development teams need to participate in user feedback sessions directly; otherwise, they don't learn the underlying needs and viewpoints of the user.

Outcomes Matter More Than Outputs

We explained that one of the dysfunctions that Agile tried to remedy was the excessive focus on big detailed up-front plans. Another was an excessive focus on artifacts.

One can describe the 1990s as a "methodology craze." Prior to the 1990s software projects were often conducted without a methodology. Instead, an experienced software engineering manager was put in charge of a group and told to build something. How they did it was up to them, and to a large degree their task was seen as an art form.

Software methodologies existed, but their use did not come to be expected until the 1990s, and so during the 1990s it was increasingly mandatory for a methodology to be chosen for any software project. One of the methodologies that became popular was the Rational Unified

Process (RUP), an iterative process. Another was adopted from the Project Management Institute (PMI), which had defined a process based on a waterfall model.

Both RUP and the PMI process were defined in terms of artifacts to be created. According to these processes, if an artifact was complete, then that step of the methodology was considered to be complete. Thus, for example, if a design document was reviewed and approved, then the design phase was considered to be complete—even though there was no evidence (other than the review) that the design would work.

Too often projects were on time and on budget, as document after document passed review, and the software was then developed as a large, single effort, until finally the software, which was checked as "complete," went into a test phase. During the test phase all kinds of problems were found, often revealing that foundational design aspects of the system were flawed, and to fix them many more months would be needed—even though the design had been signed off on as complete.

The Agile movement tried to expose this artifact-centric approach as unworkable, acknowledging the reality that it is impractical to try to design a system up front without actually building any of it to validate it. A product design cannot be known to be complete until the system is shown to work.

For this reason, the Agile Manifesto states that one should "Deliver working software frequently, from a couple of weeks to a couple of months, with a preference to the shorter timescale."

This is a huge and beneficial change in thinking. The problem is that if you are building the wrong software, then you still are not better off! Working software does not prove that you have met your business goals. It only proves that you created something—but not necessarily the *right* thing.

An effective theory of product development—software or otherwise—needs to acknowledge the value of working product features throughout development as proof that the

Agile 2 Principle: The only proof of value is a business outcome.

design of those features actually works and that those features have actually been completed; but there is even more value in demonstrating the effectiveness of those features. This can be done by deploying the features for actual users to try or use and then obtaining statistics on the level of usage or the satisfaction on those features; and also by investing in early pre-development involvement of actual users, using participatory design techniques, which we will discuss in Chapter 8.

This is not always possible. Some features require many other features to be present for any of them to be usable. For example, an airplane cannot fly without its flight control system, nor can it fly without its landing gear. So one cannot build part of an airplane and then ask actual users—passengers—"How did you like that landing gear?"

But sometimes it is possible. Etsy became famous for measuring every user interaction with its website to determine which features people liked and which they did not. Etsy used the data to decide which types of features to focus on.

Actual business outcomes should be an essential part of any Agile approach. Measurement of feature performance, usage, and lifecycle issues should be built

Agile 2 Principle: Data has strategic value.

into the system as each feature is built, and that data should be used to inform product managers and teams about the effectiveness, value, and reliability of those features. Real-time metrics and trends of all aspects of a feature—the technical performance and the business usage—should be thought of as each feature is defined and implemented. Only then can the organization achieve business agility.

There Needs to Be a Product Value Vision

A common mistake that startups make is when a technologist has an idea and invests a huge amount of effort to build and launch a product, only to find that few people want it. One of us is familiar with a security technology startup launched by a university professor. The team obtained many millions of dollars in venture capital funding and hired an experienced CEO and then spent nine months building the product, but the CEO conducted a survey and found that no one would actually buy the product. As a result, they had to scrap what they had developed and build what people said they would buy.

It is really important to know what your prospective users truly want—not what they say they want in casual conversation but what they *really* want, in other words, what they would be willing to pay for. What people usually will pay for is not what would be nice to have; what they will pay for is what they *need*. They will pay for something if it solves a pressing problem or if it satisfies a burning desire. Otherwise, they might like to have it, but they will not actually pay for it or go to trouble of obtaining it.

But sometimes people don't know what they need until they see it. For example, who among us visualized the iPhone before it was put on the market? Yet once we saw it, we knew that we needed it, or at least wanted it so much that it defined a new paradigm—the smartphone—and set the standard for what phones were and needed to be in order to compete. As Henry Ford famously said, "If I asked people what they want, they would have said a faster horse."

> **Agile 2 Principle:** Any successful initiative requires both a vision or goal and a flexible, steerable, outcome-oriented plan.

Clayton Christensen felt that knowing too much about what a user thinks they want can lead you down the wrong path. He felt that instead you need to try to understand what the user is trying to accomplish. He called this the user's "jobs to be done."[10]

Knowing the user's jobs to be done requires really understanding what the user does and why. It is not just about looking at what they do; it is about deciphering *why* they do it and then trying to envision ways to enable them to do that better, even if it means inventing new ways of doing it. That is how you crack a market open and become a market leader, rather than adding more and more incremental features that do not significantly improve things for the user.

That is why there needs to be a product vision, a concept that is informed by true understanding of what users do and why.

There Needs to Be a Technical Vision

Have you been to Manhattan in the United States? If so, did you go downtown, near the World Trade Center? Or perhaps to Greenwich Village? While you were there, did you notice the haphazard layout of the streets?

The upper part of Manhattan was planned much later, and its layout is logical—boring but easy to find your way because the street layout is consistent and follows a pattern. But the older parts are seemingly randomly arranged, because when they were created, there was no plan: each was created according to a need at the time.

Products are like that. If one approaches product design incrementally, one ends up with something like Greenwich Village, where Barrow Street and Commerce Street both merge into Barrow Street, with Commerce Street disappearing, and so one is driving on Commerce one moment and then all of a sudden is on Barrow. Or like downtown where Beaver Street begins as a westbound offshoot of Broad Street, but slightly farther up on Broad Street there is another Beaver Street as an eastbound offshoot. It makes no sense. Of course, to those living there, it is charming; but products are not like neighborhoods. They are not supposed to be charming, and people will not put up with haphazard design. A haphazard internal design is also difficult to maintain and extend, which decreases one's business agility.

This is why there needs to be a *technical* vision for how the product will work and how it fits into its ecosystem. The vision needs to be maintained over time, because markets change over time, and the

Agile 2 Principle: Technical agility and business agility are inseparable: one cannot understand one without also understanding the other.

further a product drifts from its vision, the harder it will be to maintain and extend.

Imagine a complex spreadsheet with hundreds of sheets and with formulas that link different sheets together, and imagine that when creating the spreadsheet, one had a plan for how it would be organized. But now imagine that one created it without a plan and instead created it as needs arose, always adding features incrementally. Over time it would become a haphazard mess and be almost impossible to extend.

Software is like that. Having a vision for how it all fits together is essential for being able to extend it as needs continue to evolve. In addition, since the vision for how it all should fit together changes over time as new features are added, it is necessary to continually reorganize it. Programmers call that *refactoring* the code. If one does not invest in refactoring efforts, one eventually has an unmaintainable mess.

In the Agile community there is a concept known as *emergent design*. The idea is that as more features are added to a product and the product design is continually refactored to stay organized, a de facto design takes shape—a design that could not have been anticipated at the beginning. This is certainly the case for most business software that is driven by user needs instead of technical specifications. However, the term *emergent* is misleading, because it implies that it happens on its own.

In reality, there is intention involved. If there was not a clean initial design and if refactoring efforts were not performed well, the emergent design can lose cohesion and then one cannot really say that there is a design. There is a haphazard tangle of unreliable features. Design must include intention—the intention in the technical team's collective minds for how the product will work as whole. Further, that intention must be allowed to evolve thoughtfully over time during refactoring efforts. Design does not magically emerge through piecemeal coding. It emerges if people step back and ask, "How should we change this product so that it fits together better now that we have added these additional features?"

There Needs to Be a Delivery Process Vision

For an organization whose business takes place mainly on a technology platform, that platform implements the business value delivery stream, but it is actually not the real product. The real product is the *system* that maintains the platform. Let us explain.

Look at it this way: imagine that you have a race car that you race. For the car, you have a driver, and you have a team that maintains the car. In fact, they probably built the car. The car is always changing, because the team is always tuning it to make it faster, more nimble, and more reliable. The car is secondary. The real machine—the thing that makes you competitive—is the team: the driver and the car team. That team maintains the car and operates the car.

The team (including the driver) is the real product. It is the "system" that we referred to earlier. It is what enables you to win.

In the same way, a technology platform is maintained and continuously enhanced by a team. That team, with their tools, are continuously taking new feature requests and pushing those into the marketplace. That system, consisting of the teams and their tools and infrastructure, are the real product. The user-facing products that are used to create value are the product of the system (people, tools, and infrastructure) used to maintain and enhance the user-facing products. The system is the real product.

Any product must be designed. Today's technology platforms are complex, and the systems for maintaining, running, and enhancing them are complex. That *system* must be designed, and its design must be maintained. It is a product in its own right. It is the real product that keeps you in business.

Given that the delivery system is a product and therefore needs a design, it also needs a vision. Otherwise, the design will be ad hoc and inconsistent and will not align with future needs. A vision for the delivery system needs to have a vision just as the customer-facing products need to have visions.

If your products are digital or digitally designed products, then the vision for a delivery system might include a lot of DevOps practices and tools. For example, if your products are engines, appliances, or rockets, then the vision for the delivery system might include 3D printers and solid modeling software as well as hardware simulation programs and test rigs. If your product delivery system is agile, it will enable you to design and validate product changes rapidly and create new versions to field test with very short turnaround—in other words, a short cycle time. Hardware or software, cycle time is the key.

The vision for the delivery system should therefore focus on reducing cycle time, shorter and shorter, as well as pushing validation closer and closer to the designer, so that when things are built and tested, they actually work, and time is not wasted "going back to the drawing board."

Agile 2 Principle: Technology delivery leadership must understand technology delivery.

Proactiveness—Gemba

Lean manufacturing philosophy uses a term called *Gemba* to describe a practice whereby a senior manager visits a factory floor and personally observes how the work is being done. It includes asking questions of the workers so that the manager can understand the day-to-day realities of the work, rather than relying on what management staff say.

The Agile community has adopted this practice in spirit. It often takes the form of managers locating their desks among the developers so that they can personally observe the teams and provide opportunities to observe and listen in on discussions and standups.

Gemba is an extremely important practice, because without it, a manager must rely on status reports to learn how things are going, and in a management hierarchy there is a strong incentive to *green shift* (described in Chapter 2) reports—that is, to show all progress indicators as "green" (in other words, all is okay), rather than be transparent about problems. Status received from subordinate managers also does not reveal cause-and-effect relationships. If one wants true understanding, one must discover how the work is being done firsthand.

One of us once worked for a proactive CIO. This was at a government agency, but he came from Silicon Valley, and so he had little patience for government norms and their slow pace. If he wanted to learn about something, he would walk to that office area and stroll around, and he would talk to the programmers, the testers, the tech leads. If he wanted to know something, he did not ask his staff; instead, he went straight to the source.

His organization became known as the most successful of all US government agencies at adopting Agile and DevOps methods. His name

> **Agile 2 Principle:** Good leaders are open.

was Mark Schwartz, and he later wrote some books about his experience as CIO of the US Immigration Service, notably *A Seat At the Table*.

Unfortunately, even a proactive CIO could not overcome all of the obstacles that the government imposes. One of the biggest obstacles was the government's rules about seeking the lowest-cost providers. Schwartz revolutionized the agency's procurement practices by creating contracts that allowed multiple vendors to win, each working side-by-side, with frequent renewal based on performance. The performance criteria included how Agile they were instead of specifying which features to build; but due to the lowest-cost provider criteria, quality vendors did not apply, and so the agency essentially obtained the "bottom of the barrel." In the end, some of Schwartz's improvements unraveled after he left. Still, while he was there, change was dramatic and deep, even if the immense viscosity of government norms would eventually pull some things back toward how they had been.

Managing by looking at status reports might work fine in some fields, but it does not work for product development. Elon Musk does not rely on status reports: he goes to where the work is being done, talks to the staff, and involves himself. A colleague's son worked for Tesla, reporting directly to Musk. He claims that Musk would sometimes be seen sleeping in a corner or a hallway, having stayed at the facility all night! That might be extreme, but the point is that Musk does not stay in his office or let himself be shielded from reality by layers of staff; he goes to the source and is very present where the work is being done, when he feel that that is where the challenge is.

Gemba provides transparency and makes issues concrete instead of abstract. It shows staff that you care about their reality. If you do it the right way—helping rather than micromanaging—it shows inquisitiveness and interest. It makes staff view you as a collaborator instead of an uncaring boss up on high. Gemba is a critical practice that should be core to Agile.

Organizations Need Structure

From the beginning, Agile's focus has been on the team. Extreme Programming was about how a team works together. The Agile Manifesto uses the word *team* three times, and at the end refers to the team. Scrum is a process for how a team works together.

None of these sets of ideas says much, if anything, about how multiple teams should interact or how teams should operate in a larger product ecosystem. That is why Agile ran into trouble early, when large organizations tried to "insert" Agile by converting non-Agile teams into Agile teams. There was no guidance available about how to make those teams work within the context of the larger organization.

As we explained earlier, during the early days of Agile it was common for Agilists to blame the organizations, criticizing them for "doing Agile" instead of "being Agile" but without saying what it meant for an organization to "be Agile," since the Agile literature was all about individual teams.

An early entrant to the thinking about how teams should operate within an organization—how the organization could "be Agile"—was the 2010 book by Larman and Vodde, *Practices for Scaling Lean & Agile Development*. To its credit, the book did not prescribe a methodology, but instead suggested a series of practices that could be "tried." Meanwhile, scaling frameworks popped up, and Larman and Vodde created one of them, known as Large Scale Scrum, aka LeSS.

We see nothing wrong with frameworks. To us, a framework is just a set of ideas—as things to "try", as Larman and Vodde suggested. A problem arises, however, if one views a framework as something to implement. That is, if one attempts to reproduce what a framework prescribes in one's organization, then first, one will not have gone through the learning of trying things little by little, and so one will not truly understand how the pieces of the framework interoperate and why they are there. That is critical, because all of Agile is about applying contextual judgment, so if you do not understand the reason for a practice, you will not have judgment about it, and your decisions will be wrong.

Second, no framework is complete. Every framework misses things that are important for your organization, so if you implement a framework, you are sure to be missing important things. But you will not know you are missing them, because you will not really understand the framework that you implemented.

An Agile framework is not like an COTS package installation such as SAP, where you install software, customize it, and then roll it out. An Agile framework is not like that, because you are not installing software; you are installing human processes, and those processes are not procedural steps: they are opportunities to make judgments. If your people do not understand the framework, they have no judgment about it. The processes will all be chronically broken, because the missing ingredient—the judgment that needs to be acquired over time by trying methods one by one—is not present.

The point is that the wrong way to add structure to Agile is by implementing a framework. A framework is a source of ideas and should be seen as nothing more.

You do need structure, though. If your product requires 30 teams and that product exists within a larger technology platform containing tens (or more) of interoperating products, then no team is fully autonomous, and no product is fully independent of the others. In that situation you have an immense number of interdependencies between products and components—and hence teams. If you stand back and expect them to all self-organize, you will be very disappointed.

The old hierarchical model will not work well for this. Even if it were populated with well-meaning servant leaders, a hierarchy is too inflexible.

Agile 2 Principle: Leadership models scale.

Products are not hierarchical. A product can interact with many others, and some of those might even share components, such as core microservices. The dependencies are not hierarchical. And all of those exist on technology platforms such as cloud provider networks, which need their own level of oversight, spanning all of the products; and above that, there is a level of cloud architecture and account management that impacts every product. So when issues arise, it is critical to be able to identify who is impacted and rapidly assemble those parties, and only those parties, to resolve the issue. The issues rarely fit the organizational hierarchy.

Remember that speed—agility—is a really important driver here. There is no time today to assemble a committee to plan how to resolve the issue, spending months studying it, meeting weekly, making a final recommendation, and then creating a special team to solve the problem. That is a 1980s approach, and it is too slow for today. What needs to happen today is that the right people need to be pulled into a discussion and rapidly reach a resolution, in days—if not hours or minutes!

The question is, how do the affected parties—the ones who need to be involved—get identified? If everyone is spammed all the time about every issue, you will have a population who stop paying attention.

This is where leadership above the teams comes into play. There need to be people whose job it is to watch the entire product and observe (or be told) when something is not right. Those people can in turn make judgments about who else might be affected—which other teams or products. They can then assemble the right people to work the problems.

It is a field-level decision process. And it requires having leaders on point to observe problems and take quick action—not dictatorial action (unless an instant response is needed) but collaborative yet timely action. Get the right people together. Drive a thoughtful and inclusive discussion. Push for an informed resolution. If there are competing theories and no timely consensus is possible, make an autocratic decision, *but own it*—take responsibility for it, no matter the outcome. But with good leadership and well-facilitated discussions, there will usually be consensus.

We still have not explained what kind of structure is needed above teams. We will not say. To prescribe that would be to create yet another framework. What we will say is that structure is needed and that each part of that structure needs leadership. Each structural component is, in effect, a team or a team of teams. Each component needs many if not all of the kinds of leadership that we have been talking about. A team of teams is very much a team like any other.

Some frameworks define a kind of template for structure. We feel that using a framework as an *initial mental model* for designing one's processes is a valid approach, but it should only be a *mental* starting point. One should not begin work using a framework as-defined. The most experienced leads should collaborate to define the process to start with, taking as many ideas from the framework (and other frameworks) that they think apply and adding additional ideas to address the issues that we have described here and any other issues they anticipate. Over time they should continue to improve their approach.

The Individual Matters Too

The team-centric focus of the Agile community misses an important aspect of what was defined by the Agile Manifesto. The manifesto's first value begins with "Individuals"—not "Teams." Yet the importance of the individual has been lost. The Agile community has a culture of celebrating teams and discouraging any recognition of an individual. On top of that, the emphasis on collective code ownership (an Extreme Programming idea) led to the emphasis on "generalists" who can work on anything, rather than specialists who do only one thing.

Generalists make a team effective because no one has to wait for anyone else if everyone knows how to work on any aspect of the overall work. However, seeing everyone as a generalist leads naturally to seeing everyone as equal, and that in turn leads to the idea that no one is ever recognized individually—only a team is recognized. But that view not only dehumanizes people, turning them into Taylorist machines, but it makes any kind of professional advancement impossible. How can you advance if you are seen as the same as everyone else? And where would you advance to?

Thus, the focus on teams became an extreme, whereby only teams matter but individuals do not exist. Team practices are chosen. Team stories are created. Team norms are defined. A team collectively decides. There is no room for individual difference. This is nothing short of a devaluation of the individual, even though the Agile Manifesto begins with a statement about the importance of individuals.

Another result of the devaluation of the individual is that team leads have a hard time creating individual relationships. The Agile team room's open plan layout means that there is no privacy, so a team lead cannot sit down with someone and say, "How's it going?" and expect the kind of private sharing that might occur in a private office. It is always assumed that an Agile coach or team lead coaches the team, but never does one hear about the individual.

One-on-one relationships between leaders and those who they lead are important. In *DevOps For the Modern Enterprise*, Mirco Hering writes

> *have regular one-on-one meetings with the people who report directly to you. It's very hard to manage someone as a person when she is just an anonymous entity that you only know through her work product. And no, having an open-door policy is not the same as setting up one-on-ones.*[11]

If people are all out in the open, then to have a private conversation, one might need to say, "Let's go for a walk." Making that a routine is important so that one-on-one discussion is frequent enough.

Many in the Agile community would tend to default toward public, open discussion of problems. The problem with this is that it presumes that those listening have only good intentions, which might not always be the case. An individual might need to share challenges with co-workers, and leadership intervention might be necessary; or, an individual might want to share career aspirations but not want to do so in public.

Perhaps the open team room problem will go away as a result of the inexorable shift to remote work. However, even though remote work and a globally distributed workforce are clearly on the rise, there will still be many teams that meet in person. The point is that there is a continuing need for private conversations—not as one-offs, but as a normal and ongoing activity. Open team rooms make that difficult but are merely a symptom of the devaluation of the individual in favor of the group.

Individuals matter. People are not the team they belong to. People have different talents, skills, concerns, goals, and challenges. They are not cogs in a team wheel. Teams matter, but so do individuals. A healthy Agile philosophy needs to acknowledge that loud and clear.

Trade-Offs Between Realizing Value Today and Maximizing the Ability to Produce Value Tomorrow

During the early days of the Agile movement, one of us was debating the issue of iteration planning, and the other person, who was recognized as a regional leader of Agile thought, insisted that one should not plan any further ahead than the current iteration. He was talking in the context of Extreme Programming and two-week iterations.

His contention was that anything further out was immaterial, and "you ain't gonna need it"—a phrase that is often abbreviated as "YAGNI."

This did not seem right. Yes, Agile thinking exposed the absurdity of long-term detailed planning and up-front detailed requirements: how can one know fine details far in advance? And if one precisely defines a product before any of it is built, how can one imagine how usable it will be? There is nothing like seeing a working product to crystallize one's thoughts about how it can best work.

But thinking ahead is essential, and eventually the Agile community came to realize this and now embraces what some call "rolling wave planning," in which one plans ahead but with decreasing detail further into the future and always keeping plans tentative.

The balance between short term and long term is always a tension, whether it pertains to planning, goals, or risk. Animals think only of the current moment. Humans think about the future, but some humans think further ahead than others.

There is no right answer. The future is not here. In economics, one maximizes expected present value; that is the "real options" model. But how far ahead do you project? A year? Two years? Ten years? If you plan to cash out your investments in five years, then maximizing your return over ten years does not make sense. And if you are someone who lives for today and only today, then planning for your future is not even consistent with your values and lifestyle.

The same value-based trade-offs exist for organizations, and the trade-offs for a funded startup that seeks to be acquired or go public in a few years are different than for an established company that plans to be around for the next 20 years, let alone a government agency that views itself as permanent.

Individuals Need to Develop

You might be wondering what this has to do with personal development. It has a lot to do with it, because when an organization invests in someone, they are expecting a return, and that return depends on how long the person will stay with the organization.

There is also an opportunity cost to investing in people. If someone is allowed to have lower productivity while they learn, then the organization is losing the work that that person could have done if they were put to use in an immediately productive manner rather than allowed to spend time learning.

These trade-offs are important to understand, because the Agile community has a mantra of "allow the team to fail." We have already discussed that and the rationale. It is not about failure; it is about learning. People learn from mistakes, and if mistakes are not tolerated, then people will not try to push the envelope, and they will become less inventive and will stop learning. After all, what is the point of learning something new if you are not going to be allowed to try it, for risk of not doing it perfectly?

We are not saying that one should take unwise risks. There are limits to the degree of failed experiments that an organization, or a team within the organization, can absorb. The point is that thoughtful and careful experimentation, designed to be as contained as possible, is necessary for learning and progress.

The other side of this coin is that sometimes results are needed right away, but learning takes time. Yet in a business setting, there is a tendency to view every need as immediate; no one wants to wait. If you tell a business stakeholder that a feature will take longer because a new person is "learning," they will become very anxious and say, "Why can't you put someone experienced on it?"

And they are right: they don't want to pay the cost for someone to learn. They only want to pay for professional work. But if you never give new people a chance to try new things, they will not learn, and your teams will not be able to grow in their capabilities, because the individuals will not grow, and a team is made of individuals.

Investing in people's learning is essential if you want to increase the capabilities of your teams. This means you will often have to make judgment calls about whether someone experienced should take on some work or if someone who

Agile 2 Principle:
Professional development of individuals is essential.

is inexperienced should, perhaps with an experienced person supervising or pairing with that person.

It is a judgment call, because if the work really is needed right away—for example, if the work involves fixing a defect that has brought down one's systems—then the work needs to be done by someone who can get it done right away. But if there is time, teaching people by allowing them to try new things will have the best long-term benefit—unless the new thing they are learning is something you will never use again.

Today's product teams are always learning. It is a rare case that a team takes on new work and there are not new technologies or tools. Today's tools change every year. Product developers are adept at learning new tools; it is their norm. Do not underestimate their ability to learn new tools and technologies.

Learning is not something to leave to chance. A team lead should always be looking ahead at the capabilities that might be needed in the future and making sure that people are learning and will be ready. And if there are only two people on a team who know how to do something that frequently needs doing, getting others to have experience with that is imperative; otherwise, you have a bottleneck.

People often have learning goals. If you ask them what they would like to learn, they usually have thoughts about that. This is particularly true of product developers because in their world there are new things all the time, and they want to stay up-to-date. People will invest the most effort in learning something that they autonomously decided that they want to learn; so if you ask them what they want to learn and then take steps to get them the opportunity to learn about that, they will be more committed.

Learning is not just about tools, though. People also need to learn to act on their own. The most effective team is one that can act with minimal supervision. By giving people a chance to lead, they learn to lead. People need good leadership models. They need to be shown what good leadership looks like, and they need coaching about their attempts at leadership. And some people might not be ready to lead; it is important to assess that.

Leaders need to cultivate expertise in their teams. They also need to cultivate self-leadership and autonomy: not by "throwing them into the fire" to see what happens, but by giving them more and more responsibility and coaching them out how to approach it. Prescribing or commanding them is not usually advisable, but sharing experience, pointing out things to consider, and even making some suggestions will create an understanding in the learner's mind and prepare them for being able to make their own decisions. Also, discussing outcomes afterwards in a supportive, non-critical and reflective way is extremely helpful.

In the end, people learn best if they feel that they accomplished something through their own initiative and their own ideas, rather than merely executing what someone told them to do. So any coaching should be focused on helping them to understand the situation and the risks better, rather than telling them how to approach it.

Teams Should Decide How to Work

One of the great ironies about how the Agile movement has played out is that even though one of its principles reads, ". . .Give [a team] the environment and support they need, and trust them to get the job done," Agile has come to be dominated by Scrum, which is a highly prescriptive process. Thus, while on one hand the Agile Manifesto says that teams should decide how they work, on the other hand, they are often told that they should use the Scrum process—in effect, telling them how to work.

One argument for starting with Scrum is that new teams need to learn Agile, and Scrum provides a helpful baseline; later, when teams have learned Agile, they can make adjustments to the process. However, the Scrum process is precisely defined by a document known as the Scrum Guide,[12] and there is pretty broad consensus among the Scrum community that while one is permitted to *add* to the Scrum process, one must not *change* it. Yet, adding to something does often change it. For example, if you add garlic to vanilla pudding, it is not vanilla pudding anymore. One also is not allowed to select parts of Scrum. The Scrum Guide says, "The Scrum framework, as outlined herein, is immutable. While implementing only parts of Scrum is possible, the result is not Scrum. Scrum exists only in its entirety."

In fact, a Certified Scrum Trainer (CST) once told one of us that in presenting his training materials to get his CST certification, he had to call out anything in the material that was "not Scrum"—they did not want him to supplement anything that Scrum prescribed.

In any case, *changing* the Scrum process is not permitted. If you do that, then you are "not doing Scrum," and so changing the Scrum process is strongly frowned upon. Those who strongly advocate for Scrum tend to advise that Scrum is "just a framework," yet in the same breath insist that one must use it exactly as prescribed—two contradictory messages.

The reality then is that teams are not actually allowed to choose their own way of working: management decides that teams shall use Scrum, and from that point on, the way work is done is dictated by the Scrum Guide.

The contradiction of this has been pointed out countless times by many in the Agile community; but there is inertia to things, and Scrum is so firmly established that calls to do something about the contradiction get drowned out by the countless articles about "how to do Scrum well" in myriad online sources, including Medium's Serious Scrum series. Given the large Scrum community of people who are invested in the process, there is an army of evangelists who do not want any change that might cause people to stop doing Scrum as prescribed by the Scrum Guide, because that would threaten the position and authoritativeness of Scrum—and thereby threaten what Martin Fowler has referred to as the *Agile industrial complex* that provides training and certification based on these frameworks.[13]

Scrum is a particular way to organize the workflow of a single product development team. It basically says that the main business stakeholder, which Scrum refers to as a Product Owner (PO), should maintain a prioritized list of things to get done, and that a development team works from that list in time-boxed increments—usually two weeks. At the end of each increment period, progress is assessed, and the list of work is revised. The team also reflects on its methods and makes adjustments. In the course of work, the team normally has a short daily meeting (the standup) during which they share what they are each working on and any issues. That's really it.

The Scrum process *sometimes* fits what is needed, but it fails to address a large percent of the issues that arise in product development, and so Scrum legitimately deserves some of the blame for difficulties in adopting Agile methods. The Scrum process is also problematic for Behavior-Driven Development (BDD), which favors a continuous flow. Many in the product design field feel that Scrum-like approaches disrupted product design. We will provide background on that in Chapter 8.

On the other hand, many of the issues that come up are not unique to Scrum—they are issues that come up in product development no matter what work process you use. Scrum's failing is that it says nothing about these numerous issues; yet if one tries to change Scrum to address those issues, one often gets pushback from those who feel that Scrum should be followed to the letter.

The set of complexities and issues that one encounters in the course of digital product development at scale are fairly universal, even though some issues are more prominent for some products than others. There is therefore no excuse that a development framework should entirely omit most of those myriad issues.

The problem is not *only* with Scrum. The problem is also that people treat Scrum as a process to follow rather than viewing Scrum as a source of ideas about how to get work done. The real problem is the Scrum evangelists who maintain that if one does not adhere to the Scrum Guide, then one is "not doing Scrum," and that is a bad thing. Managers are also at fault for mandating Scrum as a process for their team, instead of collaboratively defining a process tailored for the organization's circumstances and even allowing teams to define and maintain their own process, with assistance from experienced coaches.

These problems do not exist only for Scrum. Those who use the Extreme Programming process often insist that everyone must use the practice of Test-Driven Development (TDD). However, there is controversy around that method, and many people (including some of us) do not like it. The controversy is such that Martin Fowler—a highly respected thought leader of the Agile movement—organized a series of debates about the matter.[14] Yet many Agilists believe that TDD is a mandatory practice and that if you are not using TDD, then you are not "mature" in your use of Agile methods. This attitude is inflexible and treats everyone as if they are the same and assumes that what works for one person must work for all. It devalues individual agency and is antithetical to the spirit of Agile, which emphasizes trusting people to decide how to work best.

An article in *Business Matters* magazine explains that "strict procedures can start to feel like a hamster wheel, when whatever you do, you end up in the same place again and again." The article then goes on to quote Brian Knapp, the blogger at Code Career Genius, when he explains what happens when a fixed process is forced on a team and the team is not allowed to change the process when they need to: "After a while, the bad process makes a team hate working together or working at a company at all. But, you can't really fix it because the reason the process exists is to abdicate responsibility, not to improve things."[15]

Experts and Generalists Are Both Important

The Agile ideal for a team member is the generalizing specialist. This is someone who can perform most of the tasks that their team needs to perform. Such a person is "T-shaped." That is, they might have deep knowledge in one (or more) area(s), but they have enough knowledge of the other kinds of things that need to be done so that they never have to wait. They can always do what needs doing.

The problem is, not everyone fits that profile. Is someone who did not fit that pattern ill-suited for work on Agile teams? And if so, are they ill-suited to work on today's products and services? What about someone who is an expert in some area? Are they not able to work on an Agile team? Or must they have another role, such as a supporting role?

The Agile Manifesto is silent on the issue, because the Manifesto describes how a *team* should work but does not describe the characteristics of the individuals on the team. The same is true of Scrum and Extreme Programming. Yet highly capable individuals who are specialists and not generalists are common. Sometimes highly capable people are even somewhat eccentric and are not good team players and really work best on their own. What should an Agile ecosystem do with such people?

Linus Torvalds comes to mind. He was the lone creator of the Linux operating system. Today he leads Linux development, but he

Agile 2 Principle: Both specialists and generalists are valuable.

still works alone. His office is in his home—always has been—and he communicates entirely through online mechanisms. He never meets in person. He likes it that way.

These people are "lone wolves," and dismissing them is a big mistake. In the words of Reed Hastings, the founder and CEO of Netflix, "I had a choice: hire 10 to 25 average engineers, or hire one 'rock-star' and pay significantly more than what I'd pay the others, if necessary. Over the years, I've come to see that the best programmer doesn't add 10 times the value. He or she adds more like 100 times."[16]

This means that if you reject someone who is a "lone wolf" because they do not work well on a team, then you might miss out on someone who can create amazing things.

One of us had a personal experience with that. During the 1980s we worked at a startup that had a product, but the product had become a commodity, and the company needed a new product to beat others to the market. They could have assigned a team to that task, which would have taken six months. Instead, the CTO assigned a particular "rock star" programmer to it. That programmer went home and came back two months later with the product. The company beat its competitors to the market, and that product firmly established the company as an industry leader.

If you do not think this could happen today, think again. Blockchain was created not too long ago, and the evidence is that it was created by one person, even though no one knows who that person is! The popular tool "vagrant" was written by Mitchell Hashimoto in 2010, which enabled him to create Hashicorp, which now makes many popular and important cloud development tools.

But those were a decade ago. What about today? The opportunities might be in the machine learning domain. Anyone who participates in Kaggle.com competitions knows that some innovative things are happening there. All it takes is the creativity to develop a groundbreaking model and one could potentially disrupt an entire industry. The models in Kaggle are created mostly by inspired individuals—not teams.

Teams are important, but so are individuals. The lesson is to not force everyone into the same mold. Some people are team players, and some are not—*and that is not a bad thing!* It is just how they are.

If we view lone wolves as "un-Agile," we are shunning an important category of people: people who work best on their own. Perhaps the thing to do is to leverage their talents in a way that is most effective for them, giving them special projects or prototypes to create.

There are also experts who are team players but need to be on a special team, for example, a team of DevOps experts that supports product development teams to help them to set up their development and test processes; or a machine learning team that provides models to development teams to incorporate into products. The interface between these groups needs to be carefully tuned so that their interaction is optimal.

A support team might interact best with a development team by providing an individual expert to participate in the development team as an "extended team member" for some time, either on a part-time or full-time basis. That might work well because the support team is mainly sharing knowledge and advising, but not actively developing and changing their own product.

In contrast, a machine learning team is continuously refining their model. Thus, the flow of the model to the development team(s) is ongoing, and so a smooth and automated handoff process needs to be created. The process should ensure that the model is under configuration control and versioned and has gone through some known set of tests each time. There also needs to be a process for feedback when the development team discovers issues with the model, ideally as soon as possible.

Experts are a good thing, not a bad thing. Generalists are a good thing too. A robust Agile philosophy needs to account for the full range of human personality types who can bring value to a product, service, or organization.

Not All People Work or Communicate the Same

In Chapter 1 we talked about the introvert-extrovert problem. First, it is important to acknowledge that the terms *introvert* and *extrovert* are generalizations. In reality, people are on a spectrum; in fact, they are on many spectrums. Nevertheless, the term *introvert* is useful for talking about people who generally prefer non-interaction over interaction; and similarly the term *extrovert* is useful to refer to people who generally prefer interaction over noninteraction. And of course, there are people in the middle.

Mixing introverts and extroverts on a team is just as challenging as mixing people from different political parties because they are likely to have different core values or different ways of looking at things. It is no wonder that problems might arise when they try to collaborate.

When an issue arises that needs resolution, an extrovert will say, "Let's meet and talk about that." In contrast, an introvert will say, "I need to think about that. I'll get back to you."

One approach is not better than the other. In fact, as we already discussed, for a complex issue one needs a combination of approaches. For a complex issue, communication and collaboration are a process— not an event. They take place over time, through many interactions. Introverts will attempt to write down their thoughts about a complex issue. Extroverts will gather over coffee and try to talk it through. In the end, both forms of communication are beneficial. It might even be the case that the optimal way to resolve a complex issue is to always have people write down their perspective and then always talk through their areas of disagreement; but rather than dictate a method up front, such as "the best—and therefore mandated—form of communication is face-to-face," one should allow for different forms, to be decided by those involved.

What do you do if an intro-vert is leading the team? They might try to get everyone to write down their thoughts, but some people have a hard time doing that because they

Agile 2 Principle: Foster diversity of communication and diversity of working style.

are talkers. What if an extrovert leads the team? They will try to imme-diately schedule a meeting to "talk through the issue," but the introverts will cringe at that, because they have not yet had a chance to think deeply on their own, and they might also feel intimidated in a meeting with lots of natural talkers.

This is why it is essential to be sensitive to different personality types and how they best think and communicate. A successful Agile philos-ophy must provide guidance about how teams of introverts and extro-verts, working together, can successfully communicate and collaborate.

This has implications for how the team lead or facilitator should approach collaboration. Do not assume that what would work best for *you* is best for everyone on the team. Make sure that those who are writers get a chance to advance their ideas, and make sure that those who are talkers get a chance also—and that they do not out-talk the quiet thinkers. It is up to a Socratic leader to pull ideas out of the group, by asking questions, of *everyone*.

Being proactively inclusive in this way and allowing people to choose their own way to contribute to a discussion, either in writing or in a meeting, might also have the beneficial side effect of empowering indi-viduals who find themselves at a disadvantage, due to language, having a soft voice (as many women do), or simply being reserved in nature.

Focus Is as Important as Collaboration

Agile has such a strong bias for collaboration that the availability of time and space to think and focus is often greatly diminished. The Agile team room is not a place where someone can think deeply. It is possible that we have effectively dumbed down all of today's workforce by putting them all into open team room spaces.

The ability to focus and think deeply needs to be restored as a necessary core element of work. How that is achieved is open to creative ideas, but it needs to happen.

We will not get innovative products and services if people spend their day on the surface of their mind instead of delving deep into

Agile 2 Principle: Respect cognitive flow.

the problems they are working to solve and building rich mental models. As things are today, it would not be surprising if every insight occurs outside of work, on the weekend, or "in the shower" as people say. How can one have an insight about anything if there are ping-pong games going on in the background and discussions happening within earshot?

People are social. Even introverts often like being around people— as long as no one talks to them! But it is important to be able to focus when one needs to, without having to take special steps to go find a quiet spot. One's regular spot should be quiet by default; one should have to take steps to seek out social interaction. This is because people *like* being around others, so they often will not go to the trouble to isolate themselves even when they need to, because to isolate oneself is a deprivation; but people *will* invest effort to seek out social interaction.

People should not have to wear noise-canceling headsets to drown out conversations around them. They should be able to close their door and have confidence that they will not be

Agile 2 Principle: Make it easy for people to engage in uninterrupted, focused work.

interrupted until they open their door again, signaling that they are open to conversation.

The ability to focus and think deeply, when one needs to, needs to be an element of an effective Agile theory.

These issues do not change for remote meetings. A remote meeting is still a meeting. Remote interaction is still interaction. As technology gets better, remote interaction will begin to approach in-person interaction in its fidelity. If someone creates a Zoom meeting that lasts all day, "so that we can all be together," that might be a good strategy for a group that is brainstorming and in a creative mode, but it might create anxiety for some people who need to do focused work.

Pay attention to what kinds of work people need to do and how a team's unique mix of people prefer to do that kind of work, and allow for differences. Everyone needs to be productive. The strategy of how to work is something that needs to be considered on an ongoing basis, rather than simply when a team is set up.

Planning Is Important

The Agile movement was largely a reaction to big up-front project plans divided into phases and planned out at a detailed task level. By and large programmers knew that could not work for software—no experienced programmer would ever design a project like that.

The big plan approach came from other fields where it worked. For example, in building construction it is common to create a detailed up-front plan. This can work because the construction of a building is very different from software, and in fact from any digitally designed and manufactured product. Table 4.1 shows some of the ways that building construction and software development differ.

Table 4.1 Building Construction Compared to Software Development

Building Construction	Software Development
One can see progress, visually.	Progress is difficult to "see"—there is nothing to look at.
Relationships between parts are three-dimensional—only things that are physically adjacent can connect.	Relationships between parts are potentially unlimited—anything can connect to anything.
Making a change is costly.	Making a change and re-testing is cheap.
The steps are procedural.	The steps are creative.
Plans can be reused—often tens or hundreds of homes are created from a small set of designs.	No two pieces of software are the same—one never reuses exactly the same design.
Requirements tend to be in fixed categories and simple.	Requirements are unique to each application and can be complex.

To sum this up, building construction is something that benefits from detailed up-front plans, because it is a procedural activity: do this, then do that. By contrast, software development is a creative process for which there are often complex requirements, making it difficult to define them all ahead of time. The same is true for any digitally designed manufactured product.

Nonprogrammers often do not appreciate the creative nature of software development. Recall our spreadsheet analogy. Software programming is like having to create a spreadsheet containing hundreds or thousands of sheets (tabs), each with complex formulas and links between various cells on all the different sheets.

Now imagine that you are asked ahead of time to define all of the inputs, outputs, and functions at a detailed level for each sheet. And on top of that, imagine that you are told that if while implementing the spreadsheet you discover that you need to make changes to your design, that any such changes are considered to be an error on your part, for which you will be penalized, because the customer should not pay for your errors.

That is what essentially is expected when using a "waterfall" process for building software. Perhaps now you can see how unreasonable that is.

The Agile movement is a rejection of the absurdity of processes like that. Big detailed up-front plans are the central element being rejected. Unfortunately, since Agile quickly became a movement of extremes, its recommendations were often as extreme and bad as what the movement was rebelling against. Those in the Agile community who rejected the need to plan beyond the current development iteration are an example of that.

Plans are important, though—just not big, detailed plans that go far into the future. We have already described the "rolling wave" approach to planning, and that is what is needed. The idea is to plan most carefully for the current work and less carefully for future work; also—and this is crucial—expect that plans will change.

Since the Agile movement was originally about software, we have been using that context, but these ideas apply to many different kinds of endeavors for which there is variability and that cannot easily be planned out in detail—creative endeavors. Thus, any kind of unique product development, such that the product is substantially different from those that came before it, should not be planned out in detail. That is why it has always been common practice for aerospace firms to use a "skunk works" approach for prototyping state-of-the-art systems. They know that they cannot plan it out, so cutting-edge development is more like a research effort—hence the name "research and development." Kelly Johnson, the manager of Lockheed Martin's Skunk Works program, devised 14 "Rules and Practices" for such programs, and some of them read as follows:[17]

4. A simple drawing and drawing release system with great flexibility for making changes must be provided.

9. The contractor must be delegated the authority to test his final product in flight. He can and must test it in the initial stages. If he doesn't, he rapidly loses his competency to design other vehicles.

10. The Skunk Works practice of having a specification section stating clearly which important military specification items will not knowingly be complied with and reasons therefore is highly recommended.

12. There must be mutual trust between the military project organization and the contractor, the very close cooperation and liaison on a day-to-day basis. This cuts down misunderstanding and correspondence to an absolute minimum.

Number 4 is present to ensure that the large number of design changes expected will not be problematic. Number 9 emphasizes the importance of continuous testing, with the expectation that initial designs will often have problems that necessitate design changes. Number 10 is present to provide an "escape clause" that allows some standards to be bypassed in the interest of flexibility so that the focus can be on proving a design rather than meeting standards. Number 12 is present because there is often a high rate of failure of early designs, and both the project sponsor and the Skunk Works team need to behave as partners rather than adversaries.

Plans are important, but creative work cannot be planned out ahead of time in detail; sometimes even the large milestones need to change. Plans therefore need to be seen as the current best guess, and nothing more. But having a current guess is important, because without that, you cannot pick a direction for the current work—you have to be heading somewhere. The plan tells you what direction to go in.

Creating a plan also forces you to think through the sequence of events that might unfold. It alerts you to dependencies between tasks and critical paths. These are extremely important to be aware of.

The Agile community's preferred planning artifact is a *backlog*—that is, a prioritized list of things to be accomplished. A backlog is a powerful and simple mechanism, but it is not sufficient. One also needs to stay aware of critical dates, particularly external dates on which the work depends, or dates by which external parties expect to receive things.

One also needs a design—from both a user's perspective and from a technical (engineer's) perspective. A design is a kind of plan. And one needs a *process* design—a sequence, or critical path. Without that, one cannot know how much lead time is needed when one task is dependent on another.

The rolling wave approach says that the only detailed plans should be for the current work. However, even those should not be unnecessarily detailed and should be flexible. When performing repeatable work—that is, work that uses the same steps every time, such as setting up a new computer—a plan consisting of a detailed set of steps makes sense. Programming or hardware design is not like that, though: they are creative endeavors. One cannot define a set of steps.

What a programmer or designer usually finds is that if they write down their tasks for completing a given product feature, then halfway into the development of the feature they change their mind about how they need to implement it: performing the work reveals unanticipated issues and details, and so the path to completion needs to change, or they think of a better way. One also often realizes that additional steps are needed. Thus, if one writes down the actual steps to complete a product feature and compares it with the initial task plan, they will often be *very* different.

Imagine a blacksmith crafting a new piece, and presume that the piece is not a standard one such as a hammer or sword, but is a piece that the blacksmith has never made before. If the blacksmith is making a sword and has made many swords before, the blacksmith will have experiential memories of the steps it took to successfully make swords before and will be able to simply execute those steps. There is still slight variation and judgment, but the overall steps are not in doubt. But if the blacksmith is making a spade and has never made a spade before, the blacksmith will have to experiment, pulling from his or her set of techniques, trying to visualize the effect that each technique will have, while working toward the goal.

Product development is like that, except that today's tools and products are far more complex and intricate. A new product design is always different from others, and so one is inherently experimenting.

The bottom line is that plans are needed, and planning is essential. It is just that plans need to stay flexible and not be used to penalize someone for doing what is natural when developing unique products—make changes to the plan as work progresses.

Do Not Fully Commit Capacity

Business stakeholders have a tendency to view Agile teams as a product construction machine, instead of a partnership. If allowed to, business stakeholders will fill up the feature backlog far into the future, fully committing the capacity of all of the development teams out to the planning horizon.

Doing that will kill an Agile initiative.

What happens is that the teams never have time to work on their own process. It would be like a race car going around a track continuously, allowed to make a pit stop when something breaks, but never allowed to stop for routine maintenance. Initially, performance will degrade because, say, the tires got worn or the brake pads wore out. After a while, things will start to break down more frequently, until eventually the car will start falling apart and must be taken out of the race.

If a team's backlog is full out to the horizon, then the team cannot add their own items to the backlog for preventive maintenance—or for improving and trying new approaches. This is where the race car analogy fails, because the situation for product development is *even worse*. Unlike a race car, whose design does not change during a race, a product team is always striving to improve the product and improve how it does things. But improving something takes effort and time, and if the teams cannot add their own improvement items to the backlog, then they cannot improve or even maintain their work.

In the software realm, the analog of preventive maintenance is known as *reducing technical debt*. When programmers write code, they

Agile 2 Principle: Don't fully commit capacity.

often leave things slightly unfinished—loose ends—often in the form of code that is not as well organized as it could be. This is because tying up all of the loose ends is very time-consuming and has diminishing returns. They want to finish a feature because the stakeholders are asking for it. But the programmer makes a trade-off: leave a few loose ends, and mark the feature as done. Stakeholders are okay with this because they want their features, and they generally do not care if the code is inelegant underneath, as long as the feature works.

The problem is, over time the loose ends accumulate. The code starts to become disorganized as a whole. This gets worse as features are changed over time. If programmers are not allowed time to go back and reorganize the code, the code will become difficult to understand. The time required to add new features will go up. The code will start to become "brittle." As with the race car, it will start to fail with increasing frequency—it will start to fall apart.

This situation occurs no matter what process one uses, but it is more acute for Agile processes, because in an Agile process, one generally only implements what one needs to create the set of features that are currently prioritized. That means that when future features are implemented, there is often a lot of refactoring that needs to occur to accommodate those new features. The new features can often be done in a "quick and dirty" way or in an elegant way. If there is time pressure, the quick and dirty way will be chosen, with the intention to treat the internal disorganization as "technical debt" whereby the code will be improved at a later time to make it more elegant and therefore more maintainable. If no time is ever allocated to go back and make the improvements to the maintainability of the code, it can quickly become a "hairball" that is impossible to maintain and that tends to be intractably bug-ridden.

Product teams also need to invest in their own process. They need to continuously be trying new tools and trying new ways of doing their work; otherwise, they do not keep pace with the industry.

But to do these things—to keep pace and continuously improve and to perform preventive maintenance on their product and the tools and systems that they use in their work—they need to be able to carve out time and effort to do it. If there is no room for them to do that in the schedule, then the product itself will degrade over time. It will become the slowest car on the track and will start to fall apart. A point of no return can be reached, such that the car is not worth fixing.

Team-level product owners often are receptive to requests by a team to invest some effort in improving the code and the work process. It is usually at a program level where there is pressure to deliver features, according to a product rollout plan that promises certain features by a date, with the date chosen aggressively, effectively consuming all of the future capacity through the duration of the product roadmap, or at least the next major milestone.

The lesson is to not fill up the backlog or schedule in a way that prevents teams from performing their own preventive maintenance and improve their own processes and tools. They need to be given the capability to continuously improve and refine—the product's internals and their work process and tools.

Notes

1. www.allaboutlean.com/toyota-line-layout/
2. www.youtube.com/watch?v=C_hxjGgSc4w
3. www.jrothman.com/mpd/agile/2013/08/trust-agile-program-management-being-effective/
4 www.inc.com/business-insider/tech-companies-employee-turnover-average-tenure-silicon-valley.html
5. www.techrepublic.com/article/heres-why-younger-developers-cant-stand-new-programming-languages/
6. www.zdnet.com/article/python-programming-language-creator-retires-saying-its-been-an-amazing-ride/
7. vimeo.com/192637923
8. apttus.com/blog/5-key-decision-makers-in-the-sales-process/
9. www.blackbeltselling.co.uk/members/wp-content/uploads/additional-resources/lesson-6/6-5%20Decision_making_roles%20_%20Detailed%20V1_1.pdf
10. hbr.org/2016/09/know-your-customers-jobs-to-be-done
11. Mirco Hering *DevOps for the Modern Enterprise*. IT Revolution Press. Kindle Edition, p. 188.
12. www.scrumguides.org/scrum-guide.html
13. martinfowler.com/articles/agile-aus-2018.html
14. martinfowler.com/articles/is-tdd-dead/

15. www.bmmagazine.co.uk/in-business/not-all-developers-like-agile-and-here-are-5-reasons-why/
16. www.cnbc.com/2020/09/08/netflix-ceo-reed-hastings-on-high-salaries-the-best-are-easily-10x-better-than-average.html
17. www.lockheedmartin.com/en-us/who-we-are/business-areas/aeronautics/skunkworks/kelly-14-rules.html

5 Kinds of Leadership Needed

The common aspects of management that academics usually cite are planning, organizing, leading, and controlling.[1] Leading is sometimes split into directing and coordinating, because using the term *leading* is—not intending to make a pun—misleading since one can provide leadership about any of these things. For example, one can provide leadership with regard to organizing or planning.

There is also the matter of motivating people. Directing people presumes that they are motivated, but the way that you motivate people has a significant impact on their level of commitment. We discussed that in Chapters 3 and 4.

This is not a book about management theory or organization design. That said, we cannot avoid talking about those topics because they are cornerstones of agility.

We believe that the primary gap in the Agile Manifesto pertains to its treatment (or lack thereof) of the important topic of leadership. The Agile community has historically been skeptical of any kind of long-range planning, which we have discussed. For organizing, it prefers self-organization, which we have explained has merit but is too simple a model.

The Agile community also is antagonistic with regard to any notion of controlling, preferring to empower people to do their best and to learn so that control is replaced by individual judgment. We will talk about the control issue later, in the context of risk management. Our focus in this chapter is about what good leadership looks like, in a range of contexts.

We have already explained leadership issues at length, but mainly as a description of the shortcomings of current Agile ideas or how they have been used by the Agile community. It is time to stop critiquing and start giving some advice about what to actually do!

Agile 2 defines a set of principles, which are our guidance about what to do. These are intentionally not prescriptive. That is, they state things that you should consider doing but do not say how to do those things. This is intentional, because to state *how* would be to close the door on innovative ways of achieving the result. We want people to be creative in how they approach our principles.

Agile 2 also does not presume to be all-inclusive. We feel that when someone undertakes any initiative to accomplish something, they should consider the many schools of thought and not limit themselves to just one. Agile 2 is a set of ideas. There are myriad books on leadership, and we encourage the reader to seek those out and develop their own understanding of that complex topic.

Leadership is the central problem that must be solved in order to "scale" Agile. With good leadership, the other things that are needed will follow.

Scaling Agile is not about choosing a methodology. Scaling Agile is about creating the right leadership culture within your organization: defining the leadership models that you want to encourage and establishing leaders who demonstrate those models and know how to recognize the right kinds of leadership potential in others.

Which Leadership Styles Are Appropriate

In Chapter 3 we explained the different general forms of leadership defined by the Path-Goal model, including directive, achievement oriented, participative, and supportive. We also gave special mention to servant leadership (which has variants), Socratic leadership, and mission command leadership. The Agile 2 leadership principles derive from these ideas, but without prescribing a particular form of leadership.

We also explained that leadership has a direction and a focus. Outward-focused leadership is about representing and advocating for a group; inward-focused leadership is about the relationship with the group, and it can be planning, organizing, directing, coordinating, or controlling (managing risk).

This is not a book about leadership, so we are not going to try to address the topic in its full depth, but we will point out some things that Agile 2 addresses because they are important for agility.

One is that things scale better if one nurtures a group to be more autonomous. Thus, while directing people by command might get immediate results, it makes the director a bottleneck and also makes the team dependent on the director. It also tends to make people stop trying to improve things or take the initiative; and except in cult-like situations, ordering people around is terrible for morale, because no one likes to be reduced to an order taker.

That is not to say there is no place for it. The military is famous for being directive in its approach to leadership; however, what many people do not know is that many militaries, including the US

Agile 2 Principle: Provide leadership who can both empower individuals and teams, and set direction.

Military, have a strong preference for developing people to be leaders and for cultivating autonomy. The mission command model, which was conceived in a military context, is widely taught in military academies, and it is based on the idea of training troops and field commanders to act with autonomy and their own judgment. In the chaos of battle, every officer is expected to be able to change their plan based on local information to achieve the objective.

What the military knows is that when you are in the field and seconds count, you are often on your own, and you need to make your own judgments; but it also knows that if people do not stay true to the mission plan, things can fall apart. If you receive a command, you are expected to follow it to the letter; however, commanders know that their people need to also be able to make their own decisions.

So it is a balance, and the value of both direction and autonomy is recognized. The mission commander is responsible and accountable for the safety of the troops and for the mission's outcome, but good commanders know that they get the best outcomes if people are trained to be able to make their own decisions. In the words of David Marquet, writing in *Turn the Ship Around,*

> *"What you want to avoid are the systems whereby senior personnel are determining what junior personnel should be doing"* [2] *and "Empowerment: We encourage those below us to take action and support them if they make mistakes. We employ stewardship delegation, explaining what we want accomplished and allow flexibility in how it is accomplished."* [3]

There are trade-offs, however. The goal is to get the best from people: to leverage their brains, ideas, enthusiasm for the goal, and their individual inspiration and energy. You don't get that if you tamp people down and don't give them credit and don't listen to their ideas. But a leader also sometimes needs to make decisions, especially if lives or substantial property are at stake, and so a directive might be called for in some situations. Like an ace up your sleeve, don't use it too often, because you only have so many before you will start to be perceived as dictatorial.

It is not simple to cultivate leadership in people. It is not a matter of sending them to a class. Leadership is personal. People need to see it, and they need to try it.

One of the most important aspects of leadership is finding a way to get the most out of people's brains. A good leader uses Socratic methods to pull ideas out of people and get them to think. Every idea should be given a chance. No idea should be mocked. The best idea should win.

The first step is to have the right people in the room. If the room does not contain the expertise needed to achieve the objective, then there will be a long road for developing the team; but if you have a competent and experienced team, then you need to use it!

There are videos of Elon Musk leading discussions. He goes to a whiteboard and asks, "What are we going to do about this?" The discussion then becomes a set of ideas, each considered and debated, with no one talking over anyone else. Musk does not hold back in making his own suggestions, but people know that they can pick his idea apart, and he will be grateful that they found a problem with it, because his goal is that they all succeed. They know that his goal is not to look good. His goal is to get to Mars or shift the transportation industry to solar-generated electricity. It is about the goal—not about him.

That is what makes Musk a servant leader, even though he is known for being quite directive when he wants something done. People know that he is all about

Agile 2 Principle: Good leaders are open.

ideas and finding the best one, and he has shown that his judgment is good, and these things inspire people and make them trust him.

That means that people need to know that they can speak their mind. In the words of David Marquet,

"We called this 'thinking out loud'. . . So, in order to make the fewest mistakes when reporting on things, [people in a command-and-control organization] say as little as possible. This is a problem throughout the submarine force, and we worked hard to encourage the entire crew to say what they saw, thought, believed, were skeptical about, feared, worried about, and hoped for the future. In other words, all the things that don't show up in the Interior Communications Manual. We realized we didn't even have a language with which to express uncertainty and we needed to build that."[4]

If people are afraid to say what they think, their leaders will not get their staff's full brains. The leaders' viewpoints will be narrow and limited to only what they themselves already know.

Leadership at Each Level

Any permanent unit or team—we will just say *team*—within an organization is a kind of product. It provides value toward the organization's mission, it needs to be managed to ensure that it operates well, and it needs refinement over time as needs change. The "users" of a team are other teams within the organization.

These other teams are either customers or partners. If a team views itself as a service to other teams, it might have a "customer is always right" approach and attempt to satisfy the request of each customer; or, it might view other teams as partners and collaborate with them to find the best approach for them both and ideally for the organization as a whole.

Agile 2 Principle: Organizational models for structure and leadership should evolve.

The customer-oriented approach is common for overhead services, while the partner approach is more common for direct or indirect revenue-generating functions, including product development, that are funded through what are essentially revenue or investment streams for operating and improving an external customer-facing product or service.

The customer-oriented approach can operate as a cost center, which is funded through overhead, or as a profit center, which is funded by internal customers.

Despite its name, the profit center approach tends to starve the service because no customer will want to pay for improvements to the service. On the other hand, it makes sure that the internal customers only use the service when they truly need it—assuming they are not forced to use it. If they have no choice about using the service, service will likely be terrible because customers are captive. This results in the worst aspect of a cost center, which is that the service has a monopoly on what it provides and so has no incentive to improve, and also the worst of a customer-oriented approach, which is that the service has no funds to improve its service.

This is relevant because the choice of how the group is funded and incentivized has a strong impact on how it views its own mission and therefore how its leadership will behave. The motto "the [internal] customer is always right" leads to very different behavior than "we must do what is best for the organization."

Whether a team operates as a profit center or cost center or is funded through the product's revenue and investment funds, each team—and a team of teams is still a team—essentially defines a kind of product of its own. Even if the team is a "feature team" that works across many parts of a customer product along with other teams working on the same customer product, each team is still a kind of product of its own, because it performs a service: it accepts (perhaps negotiates) feature requests and implements those according to a kind of service level agreement, which might be specified by a role such as "Scrum team."

This means that each team within the organization—and again, every group at every level within the organization is a team—

Agile 2 Principle: Leadership models scale.

is a service or product team. As such, *each such team has the same kinds of leadership considerations as every other team*, and the factors and circumstances of each team should inform the leadership models that are applied.

For an actual customer-facing product (or service), the following tiers usually exist. Note we say "usually" because there are a great variety of organizations. Note that these are not a strict hierarchy. Don't memorize this list; we are just making a point.

- **Platform**—a set of products and/or services that interconnect, which enables users or higher level services to function
- **Product or service**—something that is used by multiple users or multiple other products or services
- **User product or user service**—a product or service that a *user* chooses whether to use
- **Value stream**—one or more intersecting sequences of activities that result in the production of value, from the perspective of the organization. A value stream begins with customer or user capture and goes all the way through fulfillment to the customer or user.
- **Development and delivery stream**—one or more intersecting sequences of activities that enable or create (develop and deliver) a product or service. The developed product or service then becomes part of a value stream.
- **Supporting services**—teams that assist or coach product teams or that provide governance with respect to particular areas of risk
- **Component**—something that is used by products, services, or other components and that is versioned and maintained as a unit
- **Team**—a cohesive group of people who help to maintain a component, a product, or a platform
- **Team member**—the meaning of this is obvious!

Leadership is needed for each of these, and leadership is needed for the many supporting functions that operate in parallel across these tiers. There are of course other functions within an organization beside these, but the previous list pertains to the value-generating products and services.

We are not trying to definitively identify the elements of any organization. Rather, we merely want to point out that each element needs leadership, and these elements do not form a hierarchy. Any kind of cohesive unit of activity or behavior within an organization requires leadership to maintain its cohesion and effectiveness over time and keep it pointed in the right direction, and some of these overlap.

Solving how to organize these many dimensions of leadership is the problem of organization design. We are not going to say anything about how to organize these tiers; this is not a book on organization design. Our focus is on the practices performed by the various groups that compose these tiers—the groups of people working together, whatever those groups are.

Common Types of Product Leadership That Are Needed

There are many kinds of leadership that are needed to envision, define, develop, operate, support, and maintain a complex technology product.

When we say "leadership," we do not necessarily mean directive leadership that applies authority. Authority is often a useful, and sometimes necessary, aspect of some leadership roles, while at the same time authority can be tempting to misuse and can present more problems than it solves.

Balancing the need and use of authority is an art, and often the best use of authority is subtle; so we are going to postpone talking

> **Agile 2 Principle:** A team often needs more than one leader, each of a different kind.

about it for the moment. For now, consider that leadership is any means of getting people to follow, including inspiring people and gaining their trust. Leadership is *influence*.

Every situation is different, so creating a definitive list of the kinds of leadership needed is not possible. However, some of the kinds of leadership that are usually needed are described in the following sections.

Let us reiterate that this set of kinds of leadership is not intended to be complete; it is typical for organizations that develop and offer products and services. These types of leadership are *not* intended to be roles: one person can sometimes fill more than one of these leadership needs or part of one need and part of another. Finally, some teams are so strong in the leadership abilities of their members that they do not need an explicit team lead.

Two common major divisions of leadership are *product value leadership* and *product implementation leadership*, which we will instead refer to as *product development leadership*, because that is how product implementation groups usually describe themselves.

As you read through the following section, be aware that we use the term *products* to include products and services, since a service is really a kind of virtual or on-demand product.

Here is a map of the topics we are about to discuss. Note that to be Agile, all of these functions must be "joined at the hip." None can operate in isolation. They must operate as merely distinct aspects of a single continuous process, so the leadership of these distinct aspects must find a way to collaborate effectively and continuously.

- **Leadership about product value**—the product vision and what the product does
- **Leadership about product development/implementation**—the product's creation and support
  - **Leadership about product technical design**—how the product works, internally
  - **Leadership about product development workflow**—how the teams work together, and with other stakeholders, and the workflow and human dimensions
  - **Leadership about product technical practices**—the technical aspects of the workflow
- **Individual leadership**—how individuals conduct themselves

Before we begin, let us stress that each of the following sections describes a topic for which leadership is needed in a product development organization. As we said, *these are not roles!* These are not positions to assign! An individual might provide leadership in multiple of these areas, and some areas might have leadership from multiple individuals. Instead of viewing these as roles, think of these topics by asking the question, "Do we have someone who is providing leadership over that?"

Leadership About Product Value

In an organization that develops its own products, the development teams usually refer to "the business," and when they do so, they are referring to people who define what a product must do.

In referring to the business, they mean those parts of the organization that,

- Operate the products' user-facing value stream(s)
- Via the previous point, oversee the delivery of value to users (and therefore the customer)
- Have direct access to customer acquisition (sales)people and possibly direct access to actual users or customers
- Have direct access to market analysis people
- Maintain a "go to market" strategy for the product
- Know how to assess product success, in a business sense
- Are—because of the previous items—in the best position (of all those besides users) to state what the product should do, in other words, provide product value leadership

You will see in Chapter 7 that the view that the business knows best about the product is not always true, and there are other ways of looking at the task of defining a product, but for now let's just focus on the leadership question, with regard to product definition.

As with anything that needs an outcome, leadership is essential, and there are many aspects and modes of leadership that might apply.

Product value leadership oversees the product's functionality so that leadership needs direct access to (even better, direct experience with) the user's "jobs to be done"—what the user is actually trying to get done and therefore what kind of product or product features they might need. An important reality is that the user might not know what they need until they see it. That is why product value leadership needs to not simply ask users but also needs to have a nuanced understanding of the user's world and their challenges.

There is also the matter of usability: even if you identify the user's "job to be done," the product still needs to be usable rather than cumbersome, and that is where user experience design comes into play. User experience (UX) experts are adept at defining user-friendly methods for interacting with technology products. This usually requires iterations and feedback from real users. We will talk more about that in Chapter 8.

Given that product value leadership requires attention both to (a) the user's jobs to be done and to (b) the usability of the features that support those "jobs," the question remains of what modes of leadership are suitable. For example, should there be a servant leader in charge of the product? Should they have final say authority over product features? Should they operate in a collaborative way, coordinating multiple important user communities or sponsorship stakeholders? And should there be a single product value leader—often called a *product manager*—who is accountable for the business success of the product?

There is no best answer for these questions. However, the following sections discuss some considerations.

How Visionary Is the Leader?

Should the product value leader be entrusted with defining the product vision based entirely on their own instincts? Steve Jobs was a proven visionary, but most product managers are not a Steve Jobs, and so most managers need to leverage the ideas of all their staff to have any hope of success. Still, if you have a Steve Jobs, you do not want to stifle them: people who are visionary often need to be in charge, because others do not always listen to them.

Elon Musk is a good example. He is soft spoken, and if he had not created a product of his own early in his life but instead worked for a company, he would probably be the quiet dreamer who no one pays much attention to. If you do not believe that, watch one of the Elon Musk interviews[5] by Joe Rogan and ask yourself, *If Musk were an unknown and were in a group of people, what would they say when he expressed his ideas about wanting to go to Mars?*

If you think that Musk is exceptional and that you could never do what he does, think again. Musk is in a unique position because he leveraged his smarts early and got lucky at the age of 28. But rather than be a one-hit wonder, Musk has proven that he has a highly effective way of leveraging the minds of others.

He is successful in his most famous endeavors, we believe, because of the *way he works*. He was not the one who designed SpaceX's rockets. He was not the one who designed Tesla's batteries. He is brilliant, but it was not *his* brilliant mind that created those things. What Musk did is leverage experts, bring the right people together, stimulate thinking and discussions that led to novel approaches, and then initiated action that led eventually to breakthroughs.

Examining how these people work is informative, because we all can work that way when we are given a leadership opportunity.

Vision Cannot Be Handed Off

Is a vision even needed?

Some products are built to a specification. For example, a bridge is fully designed before a single ditch is dug. However, it is often the case for hardware products—machinery, parts—that a market analysis is done and prototypes are created. Regarding whether a trustworthy design spec can be created entirely ahead of time therefore depends on the product: for a bridge, yes; for other things? Better create prototypes and see how well they work out.

The question of whether you need a vision reduces to how confident you are about the requirements. If you are extremely confident, then a vision is superfluous; but if there is doubt, then a vision guides the process of discovering the best design, informed by talking to users and trying prototypes. In that case, there needs to be a vision, in other words, a concept for why the product is needed and what makes it special; and there needs to be leadership driving that. It need not be a single inspired individual; ideally it should not be a committee. It should be one or a small number of people who "get it" with respect to what this product is for.

A common situation for a large multiteam Agile program is that a marketing group defines a product concept or feature concept and then "throws it over the wall" to a set of "product owners" who are then tasked with filling in the details. These "product owners" have to guess about what users actually need; they often have domain knowledge about the data or the business functions involved, but they do not have a real user's perspective.

In these kinds of situations, what is missing is personal investment in the actual product vision. None of the "product owners" participated in defining the product concept: it was handed to them. They did not go through the learning that accompanies market-facing trials, discussions with salespeople, and user focus groups.

Even if there need to be many individuals representing the product requirements, simply because the product is big and complex, they need to be led by someone who participated in crafting and refining the vision—someone who has talked to real users and has firsthand contact with the user's need or "job to be done." We do not necessarily mean the sales group, which might have a short-term perspective based on current products and sales targets.

Most leaders are not particularly visionary. A product manager who is in charge of a consumer product does not necessarily need to be visionary as an individual, but there still needs to be a vision. If they have a good marketing team who has an understanding of the customer's "jobs to be done" and is getting feedback on product feature ideas from real customers, then the vision can emerge through inter-actions, experiments, and analytical reflection. The product marketing and development teams need planning, coordination, and direction, but that can be done in a variety of ways. Vision does not have to come from a single "leader"—it can come from the teams through a process.

Still, as in any leadership situation, sometimes people on teams cannot agree, so a need to arbitrate a decision might arise, and someone might have to make a decision on an issue that cannot be resolved through discussion. Someone might need to say, "Trust me."

Trusting an individual's vision is risky. If someone says, "I know what people really want," do they really? If they were right once, was it luck? Or do they really have a gifted sense for what will succeed in the market?

A visionary needs people to trust them. Their vision is often coun-terintuitive. It often defies conventional wisdom. It could even be that if asked, people will say they do not want what the visionary proposes; yet when the vision becomes reality, perhaps in the form of a real prod-uct, they suddenly realize that they *do* want it. Therefore, just because people say something in a survey does not mean that you can believe what they say.

This is why test marketing is so important. Most people cannot visu-alize something they have never seen. If in 1995 you had said, "Imagine a phone without a keypad that can access the Internet," most people would have said they do not want that—that they would never use the Internet on a little phone and that any decent phone must have a physical keypad. But when people saw the iPhone, suddenly they real-ized that they *could* use it and that it was better than anything else on the market. They had to see it before they could want it.

A visionary can imagine products like that, but such visionaries are not the norm among decision-makers; product designers often have visionary ideas, since it is their job. Those ideas merely need to be tried and tested in the market.

How Competitive Is the Product Space?

If the product space is highly competitive in terms of product features, then being deeply in touch with users is imperative, so an aggressive team-based approach is needed, which continuously validates ideas and polls users; or, if you actually do have that one-in-a-million inspired product visionary, then they will guide the organization to define a winning product.

If the product is a commodity, then cost is probably more important than value, so a lean and efficient approach is probably called for, with a modest focus on user experience and most attention paid to design and development efficiency. In such a situation, having a separate product design team is probably unwarranted.

Product design is a highly creative process. Most design concepts get scrapped. Product designers need to be able to place themselves in the mindset of users and visualize those users' routines and how they might use a product. Leading such a team is like leading a fast-paced research and development team—not of scientists, but of "creatives"—people who are typically artistic and intuitive by nature rather than highly analytical. They have a sense for what looks good and what people will like. They are in touch with trends, and they know how to turn an idea into a concept drawing or mockup. They work quickly, and they collaborate well because product design involves a lot of brainstorming.

What kind of leader is effective for such a team? The team needs some level or organization, and it needs to be results-oriented, but most of all it needs someone who is a "people person," because designers tend to work as a group, and group dynamics are in play most of the day. It is not like product development, where people mostly sit quietly at their keyboards. Designers sit at their desks with their Wacom tablets or iPad Pros, but more often they are trading ideas and sketching on a whiteboard. Sometimes they build one-off prototypes using off-the-shelf parts or 3D printers, and then try it out as a group and discuss it.

They need someone who is a kind of cheerleader and coach and who will advocate for them when they have trouble coming up with an idea that satisfies marketing. One of us was once the lead graphic designer for a web startup. When you are a designer, you need to make the product manager happy. They will come and check on your design. The design lead needs to make sure that product managers do not "beat up" the design team members if the design does not meet expectations.

The design team lead also needs to make sure that the design team has sufficient access to users, is able to perform the experiments that it needs, and has an effective and bidirectional flow of ideas between them and the marketing team.

The lead of the design team also needs to run interference with the development teams. If the designs are a handoff to development and the handoff is ongoing, with new feature

---
Agile 2 Principle: Validate ideas through small, contained experiments.

---

designs provided every week or two, there needs to be a coordination process so that feedback is received from the development teams. The development teams should be asking if they can participate to some degree in user discussions. The teams will have to figure out how that will work. The design lead will need to coordinate when there are hiccups in the process and help to resolve them so that the design team is not sidetracked by those issues and can focus on designing.

Leadership About Product Implementation

We explained that in an organization that develops its own products and services, the development teams usually refer to the product definition team as "the business." The people in the business have a term for the technology implementers: "tech." We thus have a dichotomy: business and tech.

These are like two halves of a brain, each incomplete without the other, yet separated by a chasm and bridged through various interactions, analogous to the extremely active and high-bandwidth corpus callosum in the human brain.

Just as *business* is an oversimplification of what the business actually does (strategy, sales, marketing, business operations, advertising, product design, finance. . .), the term *tech* is an oversimplification for what the technology side of the organization does.

Tech includes product implementation, but it also includes technology strategy, research, infrastructure, tooling support, and technical product operations. Those are all definitely within the scope of Agile, since Agile ideas have successfully been applied to many kinds of group endeavor. We need to be careful, though, because research is very different from development; and infrastructure operations is very different from either of those. Tooling support and technical product operations also differ. Let's first look at product development leadership.

There are typically three dimensions to the implementation of a product.

- The product's internal design. For a technology product, this is the product's architecture and technical design, which—for an Agile approach—will be created at a high level and then be refined over time.
- The human elements of the development and delivery *process*— who does what, when they do it, etc.
- The technical elements of the development and delivery process

Each of these dimensions needs leadership. Think of each as an aspect of product implementation, or a set of concerns and issues that need to be shepherded.

Leadership About the Product's Technical Design

The first dimension—the internal design of the product—is generally handled through a collaboration of technical stakeholders. For a complex product, there will likely be someone who is designated as a technical architect for the product. This role is a thought leadership role. Its purpose is to ensure cohesion and consistency within the product and spanning the product and other products within the organization's business platform.

As a thought leader, an architect should not act autocratically; they should behave in an inquisitive manner, soliciting ideas and sharing their own ideas. Most architects have authority over technical design decisions, so the final decision about a technical issue is theirs; but if they want the teams to feel invested in the design, they should have open Socratic discussions with the teams about design issues so that all ideas are given a chance and the best idea wins.

Leadership About Product Development Workflow

The second leadership dimension in the preceding list, the development and delivery *human process*, pertains to the workflow, so to speak—that is, how is the work "chunked," how is it prioritized, how is the decision of who performs work made, and so on. Many product development teams use the Scrum process for their workflow. Many others use the Kanban process instead. Others use another or use one that they devise themselves.

In the Scrum process, the role of Scrum Master is the leader of a team. A Scrum Master is supposed to use servant leadership. In fact, the role carries no authority of any kind. It is a role purely of advocacy and coaching. The role's definition has changed dramatically over the years and still seems to be changing at the same pace,[6] so we will not try to define it here more than we have. The role is defined by the Scrum Guide, which can be found on the Internet.[7]

Scrum aside, a team needs process leadership. That leadership can come from the team itself, or it can come from a designated team lead. In either case, the team should have a lot of say about how it does its work.

Agile 2 Principle: Self-organization and autonomy are aspirations and should be given according to capability.

The team's process leadership should therefore be flexible, and the team should be empowered to largely define its own process, while considering that if it diverges from norms that the organization has, it might have to explain why. It also needs to meet constraints pertaining to organization risk management and adhere to organization-wide standards pertaining to work process, although we would argue that such standards should be flexible and only what is absolutely necessary.

A team also needs an organizer. If a team has a process lead, they might provide organizing leadership, or someone else can. Keeping things organized is a full-time job and is a major distraction for a team, and if someone can be responsible for making sure that action items are tracked, issues are discussed and resolved, decisions are published on the team's site or elsewhere, required external communications are published or sent, and so on, that relieves the team from having to think about those things.

It also helps if there is someone who is able to inspire the team—to rally them and maintain morale, or what some Agile coaches call *team happiness*. A process lead often is perfect for that task and helps them to do their best by encouraging them, inspiring them, advocating for them, and coaching them.

Those providing process leadership should take an interest in the team's work and how the team members do their work—including the technical side. Process leads need not be expert in the technical side, or even able to do that work themselves, but they should take an interest in it and develop an understanding of the various technical practices.

A process lead should be watching for potential process, behavioral, scheduling, or morale problems of any kind. A lead should stimulate and facilitate discussions about possible gaps or issues; share insights, experience, and viewpoints; help the team to make decisions by consensus; and when that is not possible, facilitate escalation to a decision-maker—or, if the team lead has decision authority, make the decision and take responsibility for the decision.

Leadership About Technical Product Development Practices

The third dimension pertains to the technical tooling and automation of work steps. Product development consists of design, implementation, and validation. Design is deciding how something will work, implementation is making something that adheres to the design, and validation is checking that it actually works. Validation usually mostly consists of running tests.

In an Agile process, one does a little of each of these steps, again and again for each feature until the feature works or is satisfactory, and adds a feature at a time until one has a feature set that is useful. One then repeats and keeps adding feature sets.

In an Acceptance Test-Driven Development (ATDD) approach, implementation includes defining automated acceptance tests for the working product, as well as creating the working feature. That way, the acceptance tests can be run as the developer implements the feature. When the acceptance tests finally all pass, the feature is done—by definition.

An important issue is how one knows if the acceptance tests—the validation suite—is sufficient. We will not get into that yet, but it is something that is often overlooked. We will say more about this in Chapter 9, when we talk about continuous delivery as a technique for real-time risk management.

For today's products, there is an enormous amount of automation involved. The tests are mostly or completely automated; that means that creating those tests is a significant body of work that needs to be performed throughout the development process. Also, to test a product, one must simulate the environment in which the product will be used. Creating that simulated environment is a substantial amount of work too.

A common approach today is to define the environment as a template so that a test environment can be created whenever tests need to be run; this eliminates contention for the test environment, since one creates an environment whenever one needs one. It is also common to deploy the product into a real usage (production) environment and let some actual (or simulated) users use it in a live setting. All of those steps need to be automated to ensure that they are controlled and repeatable. That way, tests can be run any time there is a new version of the product. Modern product development systems produce a stream of "potentially deployable" product versions, so deployment to a simulated (or real) environment for testing, followed by a test run, is continuous.

Creating the tooling and automation is a complex endeavor; it is on par with the complexity and level of effort of the product itself. In effect, the tooling is the "product that creates the product." It therefore requires technical leadership to make sure that it is designed, implemented, validated, and continuously improved; and the design needs to maintain cohesion and consistency with practices that have been standardized within the organization. Just as for the product's technical architecture, the leadership needed is thought leadership: Socratic, perhaps with authority, but at least with a great deal of influence.

Individual Leadership

Every team member should show leadership with regard to their participation on a team and in the organization. This means being proactive in learning what is needed to perform the tasks of the team; soliciting feedback on their performance; expressing ideas for improving the work and how the work is done; voicing concerns when concerns become apparent; reaching out to other team members to help them when they need it; and being considerate of other team members by respecting their way of working, which might be different from one's own.

Individual leadership is something that needs to be cultivated and inspired. It also needs to be explicitly stated. Some people are

> **Agile 2 Principle:** Professional development of individuals is essential.

accustomed to waiting to be told what to do. A team lead needs to tell the team members that they are responsible for outcomes—not for doing what they are told. As submarine captain David Marquet put it, when speaking to his crew,

> *"So here's what we are going to do. You are all going to monitor your own departments and whatever is due. You are responsible, not me and not the XO, for getting it done."*[8]

Marquet stopped monitoring what everyone was doing. Instead, he monitored outcomes. He made his crew responsible for figuring out what they needed to do in order to achieve the outcomes and stopped looking over their shoulders, and he gave them an explicit heads-up about that, because it was not what they were accustomed to.

Everyone is their own leader. Some are better at leading themselves than others. Self-leadership benefits from being able to organize, as well as being able to inspire oneself, to create the needed motivation.

Many people find it useful to have a mentor or someone to inspire us—someone who has the attributes that we would like to have in ourselves. They provide a model that we can strive to emulate.

High-Risk Products

How much risk is there for your organization's products? For example, is the organization a government service? Government organizations usually put higher value on reliability, security, and compliance with rules than they do on new features or usability. In that case, a user experience team might be less critical, although it still might be valuable.

On the other hand, if the government agency is trying to get more people to use its online service, then the agency might actually place a high value on usability and a jobs-to-be-done perspective.

In a similar vein, if the product has high mission risk pertaining to life, safety, or large amounts of money, such as for aircraft, spacecraft, or financial services, then accountability is probably essential, and so the leadership approach probably needs to have an individual with personal fiduciary or other accountability.

Accountability does not ensure that risk is well managed, however. The Challenger disaster taught us that lesson. In the words of Richard Feynman,

"...*engineering often cannot be done fast enough to keep up with the expectations of originally conservative certification criteria designed to guarantee a very safe vehicle. In these situations, subtly, and often with apparently logical arguments, the criteria are altered so that flights may still be certified in time. They therefore fly in a relatively unsafe condition, with a chance of failure of the order of a percent (it is difficult to be more accurate).*

"*Official management, on the other hand, claims to believe the probability of failure is a thousand times less. One reason for this may be an attempt to assure the government of NASA perfection and success in order to ensure the supply of funds. The other may be that they sincerely believed it to be true, demonstrating an almost incredible lack of communication between themselves and their working engineers.*"[9]

In other words, management tends to suffer both from a gambler's fallacy and from optimism bias. Management judges that since a failure has not happened yet, the chance must be small; also, extreme external pressure to achieve success causes managers to drastically underestimate the chance that something will go wrong—even if their own staff is warning them.

This presents a strong argument for collective decision-making when there is a lot at stake. A single accountable person might be so afraid of a negative outcome that they delude themselves about its likelihood.

This presents a dilemma, because on the one hand, a high mission risk makes accountability important because if no one is accountable, no one is "on point" to watch and make sure that every risk is being managed. On the other hand, that individual will be inclined toward cognitive biases that ignore actual risk.

We do not want to prescribe a solution for this dilemma. The dilemma is a result of inherently opposing concerns: the need to prioritize that any chance of failure is addressed and err on the side of caution, and the need to make sure that progress is made and dates and times are met. It is like the "mutual assured destruction" system requirement for nuclear arms that is guaranteed to never be triggered by mistake, and yet the system must be guaranteed to be triggered on a moment's notice and with complete certainty if the need arises—these are inherently opposing goals.

The Toyota Production System took the approach of allowing any assembly line worker to stop the production line. This removed from management the responsibility of responding to a problem. Any worker could, and was expected to, although it seems that there was also pressure from other workers to not actually stop the line.

The risk of an unlikely but catastrophic failure is sometimes called a *black swan event*. That metaphor applies because black swans are rare, but they exist. The Challenger disaster might have been a black swan event, although Richard Feynman actually seemed to feel that a failure of that magnitude was actually to be expected—the warning signs were there—and that the real problem was that management did not listen to its engineers and chose to underestimate the likelihood of a failure.

The key thing is that a failure would be catastrophic; it would be "game over." If there are non-negligible failure scenarios that are game over, it makes sense to empower every person on the team to say "stop!" Odds are if someone says "stop," they are wrong, but better safe than sorry, because if they are right, it is game over.

The challenge, though, is how do you balance allowing anyone to say "stop" with making sure that some particularly reactive individuals do not say "stop" too often? As we said, at Toyota, workers who stop the line are sometimes subject to admonition from others. Thus, it is not only managers who might react badly to someone saying "stop"—other workers might as well. Groups of people are known for *groupthink*: the tendency of a group to gang up on someone who is acting outside of the norm. Thus, the personal risk of saying "stop" is high.

The solution needs to be objective, not subjective. Saying "stop" needs to trigger an analytical process, which evaluates the claim that there is a problem. It should not be up to leadership to decide, because they are just as vulnerable to pressure as well. If you allow leadership to arbitrate when someone says "stop," you might as well just allow leadership to manage all risk.

Thus, the process for evaluating a "stop" must be clear and not easily manipulated. Note that if one allows a manager to make subjective judgments about large black swan risks, one is essentially trusting that manager with the same kind of foresight that one might give a visionary like Steve Jobs, but such people are rare, so unless the manager is particularly visionary, the manager should not be trusted to make subjective judgments about black swan risks. Instead, an objective process is needed. When someone says "stop," it means that they feel that the current risk mitigation systems have failed, and they are saying, "Our process is broken. We need someone to look at the *process*, objectively and scientifically."

Research and Innovation Leadership

Research and development are often lumped together, but they are different activities. One of us (Cliff) was once a research assistant on a team trying to build an "ion lens" for the purpose of fabricating extremely small nanoscale structures. There were no task lists; there were no schedules. There could not be, because we were not entirely sure how we were going to make it work.

There were people charged with researching different aspects of the problem. Cliff's job was to create the software for modeling the ion lens. But there was physics that was needed that the team did not have, and others were trying to derive and discover that physics. There was both theoretical and experimental work involved. The team had a research grant to try to solve the problem, but there was not a specific design that was being implemented, because they did not yet know how to achieve the goal.

The process of research needs to be much more loosely structured than the process of development. When you develop something, you are confident that you know how to do it. You might need to create a design, but you have the knowledge and the skills; you merely need to apply them. If you fail, it is through error, rather than lack of knowledge.

Although that is not entirely true. In software development, it is common for programmers to use new tools—new APIs, new software libraries, new third-party components. That makes development a little more like research. If you need to use something you have never used before, then you do not really know if you can do it. There is uncertainty: you cannot be sure that you can figure out how to use the new thing. You might not be able to get it to work.

Similarly, research does actually have aspects of development. In the ion lens project, Cliff's task was to devise software using known physics. There was no uncertainty; it was a matter of designing code that would work. It was an engineering task, not a research task, even though the task existed in the context of a research project.

So, the line between research and development is gray. Research usually includes some development—perhaps a lot. Development often includes some research, but if too much research is involved, then it is not development anymore; it is prototyping.

Leading a research team, or a prototyping team, requires someone who is sensitive to the uncertainties of research. It also requires deep knowledge of the subject matter. This is necessary to be able to participate in conversations about issues, as well as plan and identify risks. An open and Socratic approach will best utilize the minds of the research team. At the same time, research can easily become unproductive, and an effective research lead needs to be able to keep people focused on a goal.

It is necessary to learn how each person works to be able to stay connected with their progress. Some research tasks involve mostly thinking; others require experiments. An individual researcher tends to be either a theorist or an experimentalist,[10] and it is important to understand the needs of each, because both are needed.

Agility in research requires being ready to pivot. If an approach does not pan out, one must judge the eventual likelihood of success, because sometimes the most successful approaches are the ones that are difficult to achieve; it could be that an approach looks unpromising, but through dogged persistence, it will eventually yield and change the paradigm. That was the case for the discovery of polonium and radium, and it was both observation and theoretical insight that led Maria Skłodowska Curie to persevere.

Agility also requires making good judgments about which lines of inquiry are most likely to succeed. That is not too unlike product vision leadership. If someone has particularly prophetic insight, they might be able to select the directions that will pan out. On the other hand, it is highly likely that the members of the research team will each have their own judgments. A perfect process for picking the right approach cannot be designed. There is a great deal of luck involved. Thoughtful and open discussion are useful for clarifying one's vision and the options that are available, but in the end, a judgment must be made.

Leading a research team also requires a great deal of advocacy. In the case of Curie, she was her own advocate, and her husband advocated for her work too. Obtaining research facilities was a great struggle. The fact that she was a woman scientist at a time when women were regarded as without standing in science was a great handicap, but results spoke for themselves, and eventually she received two Nobel prizes.

Operational Leadership

To be an operational leader, one must be an organizer and be someone who views everything as a system. One must think end to end and be interested in metrics such as cycle time and mean time to repair. One must be a good listener, because team members will have a lot of thoughts about how to improve their processes.

An operational leader is overseeing a kind of assembly line. Lean principles strongly apply. One must "Gemba walk" and stay aware of how the work is being done, rather than staying in one's office and looking only at metrics. One must have thoughtful discussions with staff, and one must empower staff to try things that might improve the workflow. One must also be an advocate so that important resources such as test environment capabilities can be obtained and so that regression tests are created for each new product feature and are maintained over time.

What about DevOps? Doesn't that say that the Development function and Operations function need to work together? Well, kind of.

The trend today is for product developers to operate and support their own product. This is a significant change from tradition, in which product developers handed off a completed product to a separate operations and support team.

Teams that develop and support their own product *still often need an operations group to help them*, but such an operations group is focused on providing supporting services, such as the core technology platform, as well as defining standards. Each group needs leadership.

Development teams should have a great deal of input into defining standards. Thus, the function of Operations is effectively shared between the product teams, which support their product, and an Operations support team that provides a set of services to make operational support easier. Both Development and Operations need to collaborate to maintain (and keep updated) an effective set of standards that are then overseen by Operations but used by Development.

The Operations support team will likely also actually operate core infrastructure so that development teams have a reliable platform on which to develop. That means there should be SLAs for that infrastructure. The Operations group will view themselves as providing a service, but they also view themselves as helping to deliver the products. They need to have a holistic view, spanning all of the organization's products. They also need funds to stay current and maintain the core platform, continue to maintain standards, and respond to requests for help from product development teams. Their leadership needs to balance all of those needs.

From an organization-wide operations perspective, agility translates into being able to respond quickly. For example, if there is a production incident, how rapidly can the problem be fixed? The traditional approach is to define tiers of support such as tier 1, tier 2, and tier 3. Tier 1 is the group of people who receive a call when there is a problem. They are a customer service function. If they cannot handle the issue, then they pass it to tier 2, which is a dedicated support team. If that team cannot handle it, they pass the issue to the product development teams—tier 3. The trend today is to dispense with tier 2.

Let's assume that there is an incident, and tier 1 has determined that there might be a problem with the systems, so they pass the incident to tier 3 (the teams that develop and maintain the product). To be able to respond to the incident, one must be able to do the following steps:

1. Reproduce the problem so that one can "see" it occur.
2. Trace the source of the problem to its initial "smoking gun"— the first point at which the anomalous behavior can be found in the logs.
3. Design and implement a fix.
4. Run tests in a test environment to determine if the fix solved the problem.
5. Deploy the fix to production, exposing the fix to actual users.

To do these things, one essentially goes through all of the steps that an Agile developer goes through when implementing a feature. Just view the problem as a feature to implement. By implementing a fix, you implement the feature.

In other words, fixing production problems is the same process as feature development. It therefore needs to use the same workflow and tooling setup. This means that your operational agility is exactly equal to your feature development agility.

We will see in Chapter 9 that development agility is contingent on a great deal of automation and the ability to create production-like test environments, as well as the availability of a high coverage regression test suite. Without those things, agility is reduced. Operational agility therefore requires all those things.

Operational agility is measured by metrics such as the mean time to devise, validate, and deploy a fix. In other words, the measure is the time that it takes to fix a problem from the time that the problem is first reported. This is often called *mean time to repair*, or something similar.

To achieve operational agility in that sense, one needs to reduce the feature cycle time and make sure that development teams have the full support of the Operations group when there are crises. Instead of "take a number," the relationship needs to be, "How can we help?" This means that Operations needs to be "joined at the hip" with Development and have an intimate knowledge of how developers do their work and what their needs are. The relationship needs to be highly supportive and win-win.

People want a product to work first and foremost; they do not want to have to call support. But when they do have problems, they want their problem fixed—the sooner the better.

Leadership and Accountability

The Agile community's preference for not having an explicit team lead led inevitably to the idea that no one is individually accountable; rather, the entire team is accountable. The theory is that if the team succeeds, it is due to everyone's collective effort. This is much like the collective farm idea of Soviet Russia: the workers all benefit from their collective effort, and no one benefits any more than anyone else. Work on a farm is work for the collective.

Another analogy is a sports team. No one on a sports team wins; only the team wins (or loses). Team spirit is about acting for the team. For example, if you think that you might have a shot at making a goal, but a teammate is in a better position, you sacrifice your opportunity and pass to your teammate, because that maximizes the chance that *someone* on the team will make a goal, even if it is not you.

Product development is not like sports, however. One crucial difference is that people on a product team have individual aspirations.

For example, someone might want to learn more about something because it is in their career plan—not because it is best for the team.

Another difference is that product development is a craft, and people have pride in workmanship. In sports, if you make a shot and the shot reaches its goal, you don't really care how the shot looks afterward—it was a shot, and it made it. You might reflect on your technique, but the shot itself is secondary, as long as it made the goal.

In contrast, the work of a product developer is an artifact that lives on. That means that people care about it, just as an artist cares about their art. In a collective farm, no one can put their name on their plot, because no one has a plot. Every plot belongs to everyone—and therefore to no one—and so the plots suffer from the "tragedy of the commons," whereby no one takes any pride in any plot and so the plots tend to be poorly managed.

Being accountable means that your name is on something. It means that when someone looks at the work, they identify it as the work of someone in particular. Individual contributors—team members—are viewed in terms of the quality of their contribution to the work.

If the accountable person is the leader of a group that produced the work, then accountability means that the leader is viewed in terms of the effectiveness of their leadership, rather than specifically the quality of the work. What matters is that goals were met.

All this will sound like a "bad smell" to many Agilists. They will want to insist that no one should be individually recognized for specific code, and that if the team succeeds, it is not because of the leader, but it is because of the team. And they are right: the team does the work and deserves credit if they succeed. But the leader deserves credit too, of a different kind. Leadership is work, and it can be good work or bad work.

Denying a leader credit for their good leadership is unfair and illogical.

Collective recognition for a job well done is a valuable practice because it honors the team. But individuals need to be able to have pride in their individual contribution, even if it was improved upon by someone else. Individuals also need to be recognized, and leaders are individuals.

There is another side to accountability, having to do with relationship building. For an organization to scale, teams or organizational units need to have spokespeople. That is, when the leader of one unit or team needs to speak to another team, it is impractical to speak to the whole team. That would disrupt the team's work, and what if they need to speak to several teams at once? That would disrupt several teams. It is more practical to speak to someone who can speak on a team's (or business unit's) behalf.

One approach is for a team to nominate someone to speak for them. The nomination can even be temporary. For example, the spokesperson role could rotate among the team members. But then when someone wants to speak to the team, how do they find out who the current spokesperson is? Do they call someone on a team and ask? Do they go to the team's website and look for a banner saying, "The current spokesperson is <Sally>"? Or do they stroll over to that team's building and floor and ask someone on the team?

And what if they need to talk to several teams? This quickly becomes logistically difficult.

Perhaps there could be a team secretary role who sets up meetings for the team and always assigns the current spokesperson. That might work logistically. But there is another problem: continuity.

When someone meets with a team spokesperson, that person needs to remember the conversation that they had last time; but if it is a different person, they will not. They will only remember what was mentioned to them when the prior spokesperson briefed them or shared notes from the meeting. So immediately we have a potential source of miscommunication.

This has an impact on relationship building. If every time you meet with the spokesperson for another team (or unit), and it is someone else, you will have a hard time building a relationship with that team's lead. By "relationship" we mean a relationship of trust and understanding. Just as teams go through forming, storming, and norming, so do people who negotiate about cross-team issues. If the person is always different, the forming, storming, norming process will be hobbled.

If there is no relationship of understanding and trust between representatives of different teams (or units), then making agreements is much harder. One cannot say, "We will make sure that your concerns are taken into account—trust me." The other person will not trust, because they have no history with you. Of course, you could take notes and record that you committed to something, but that reduces collaboration to an impersonal level, in effect introducing contracts into everyday agreements between team representatives, and contracts are something we would like to avoid.

This is why individual accountability for leaders is important. It enables them to have standing: they stand behind their commitments because they are accountable for those commitments, and that enables them to make promises and build trust with other leaders.

To remove accountability from leaders is to remove trust relationships from among the levels of management, which is not workable unless we want a Soviet-style system that is governed by rules and ultimately fear that one will be found to be in violation of a written commitment.

Any Leader

There are common traits that leaders often share. Two leaders can both be effective but have a different mix of useful leadership traits. Some of the traits that are useful for most forms of leadership are as follows:

- **They are able to align us and unite us.** By so doing, they bring us together for a common purpose to achieve a worthwhile goal.
- **They engender trust among their followers.** So, their followers are willing to give their leaders power, authority, and resources to achieve the goal.
- **They are capable of negotiating with powerful groups on behalf of their followers.** Their followers trust the leader to represent the followers' interests and to make the right decisions for them. The leader is able to develop trust in those who they negotiate with, and that the leader will be true to their word.
- **They know that they don't know all the answers.** They explore each situation, ask for advice, and listen to people. They conduct experiments and look ahead.
- **They know that they cannot do everything themselves.** They ask others to help. They know how to recruit people, inspire them, listen to them, and delegate to them.
- **They inspire those who report to them to become leaders as well.** They explain the situation and the mission, and they ask their people to develop a plan to achieve it, rather than defining the plan and issuing orders.
- **They give their people the authority and resources they need to achieve the goal they have agreed to achieve.**

- **They continuously assess people's abilities and give measured trust in people** to make decisions quickly and autonomously based on local factors. They know the world is uncertain and constantly changing and know that the people close to a situation are in the best position to make the right decision to achieve the goal. They take action to make sure that people are being trained and their skills are improving all the time to meet future needs. This is the reason for Agile 2 principle 10.9, "Professional development of individuals is essential."
- **They demonstrate the behavior they are asking for from others**. Leaders lose our trust when their words and actions diverge.
- **They define and exhibit the leadership model of their organization through their words and deeds.**

An Outside Person: A Sketch

To use Drucker's concept of an outside person, meaning a leader who is the face of the group, team, or organization, there are qualities that an outside person will benefit from having. These include the following:

- **They inspire us.** They tell stories of villains and heroes, a dark present and a bright future, of obstacles to overcome and a plan to overcome them.
- **They are goal-oriented visionaries.** They see far with respect to the future of the group or organization. They go first—they are pioneering. They paint a picture of a better world and show us how to get there.

Their vision might be customer-focused, or it might be stakeholder-focused. For example, Warren Buffet is clearly a visionary with respect to how Berkshire Hathaway manages its investments, but Berkshire Hathaway is a holding company that buys and sells companies. It is not customer focused—it is investor focused. The vision that is needed to guide Berkshire Hathaway is a vision of how to manage an investment portfolio—not how to please customers.

One could say that Berkshire Hathaway's investors are "customers" and that would be a valid perspective, but the point is that there are often many kinds of stakeholder who are important, and pleasing stakeholders is sometimes quite different from creating a snazzy consumer product.

In addition, a person who is the face of an organization or team benefits from the other leadership traits that we have mentioned.

An Inside Person: A Sketch

The concept of an inside person refers to someone who is accepted by the people doing the work as "legitimate" as a leader of the group—it is someone who people admire or trust or feel is competent to make big decisions. This is often a CTO, a chief scientist, an original founder (like Steve Wozniac), or someone who has done impressive things that inspire others, but from an internal perspective pertaining to the work that the people in the organization perform. They are often an expert in that work, but they can also simply be charismatic and someone who others relate to.

An inside person often benefits from the following traits:

- **They are goal-oriented visionaries.** They see far with respect to the way that the work is done or the subject matter or the work and its future evolution. They go first, in that they are early adopters by nature. They paint a picture of a better world and show us how to get there.
- **They share the benefits of success with their people.** They fulfill the promises they have made to them. They want others in the organization to be fulfilled and not be taken advantage of. They would not stand for unfair dilution of people's shares or elimination options in a merger. Leaders are accountable to their people.
- **They inspire us.** They tell stories of villains and heroes, recounting stories of bad actors in the business or technology domain and painting a picture of what ethical behavior should look like.
- **They teach other people how to be leaders.** They act as mentors, and help others to grow professionally.

A Person of Action: A Sketch

A person of action is an organizer—someone who focuses on execution to "gets things done"—not necessarily by doing it themselves but by leveraging the people and resources that are available.

Organizers benefit from the following leadership traits:

- **They have their finger on the pulse.** They understand the problems their teams are experiencing. They understand them logically and emotionally. They understand people's challenges, pain, fears, motivations, and desires.

- **They are deeply involved in the work.** They ask people what is going on, and they help them find the way. They set goals and constraints but not detailed solutions. They value deep and open discussion. They are inquisitive and do not get emotional when others disagree with them. One can discuss things with them honestly. They help the best idea to win.

- **They move things along.** They are sensitive to time and the need to resolve issues in a timely manner. They are good at prioritizing and balancing many things.

- **They coordinate.** They keep people informed, but only as appropriate. They are always watching for a "dropped ball."

- **They understand what we need, and they explain how we can get it.** They have a solution and a plan that makes sense that seems achievable, but they first ask others for a plan. They are good at driving discussion toward resolution and an actionable plan.

- **They make good decisions that solve problems.** When a decision is needed, they are able to make it decisively and transparently, and they take responsibility for the outcome; or, they escalate it and press for timely resolution.

- **They teach other people how to be leaders.** They ask people to describe a problem, the solution, and the outcome and to ask for the power and resources they need to achieve it. They grow the leaders that will replace them in the future. Agile 2 principle

10.9, "Professional development of individuals is essential," includes leadership development. People need to be coached on the leadership models that the organization prefers and need to be given chances to lead so that new leaders are created over time.

- **They help people to organize themselves to solve problems.** They coach, mentor, train, and support people. They understand and improve group dynamics. They make us better people.
- **They are systems thinkers.** They look at problems holistically and understand structure, stability, trade-offs, processes, tools, and budgets.

A Thought Leader

Individuals within a group often become known to be thought leaders on various topics. This happens when their ideas come to be recognized as effective or insightful.

Becoming a thought leader requires being heard. Often it is the soft-spoken people who have good ideas that no one knows about. Such people do not become thought leaders because their ideas do not get heard and discussed by others. For a soft-spoken or timid person to be recognized as a thought leader, they often need to be given some authority or designated leadership in the class of issues in which they are expected to lead.

To be a thought leader, it is beneficial to have the following traits:

- **They are able to articulate their ideas verbally.** They "keep the floor" long enough to make their case.
- **They know how to decompose their ideas into consumable parts.** As a result, each part can be explained individually, and over time all the parts connected.
- **They are able to communicate in multiple different ways.** They thereby tailor their communication to the listener or reader.
- **They demonstrate the ideas or behaviors they are asking for from others.** Thought leaders lose our trust when their words and actions diverge.

Notes

1. open.lib.umn.edu/principlesmanagement/chapter/1-5-planning-organizing-leading-and-controlling-2/
2. L. David Marquet. *Turn the Ship Around!* Penguin Publishing Group. Kindle Edition, p. 98.
3. L. David Marquet. *Turn the Ship Around!* Penguin Publishing Group. Kindle Edition, p. 181.
4. L. David Marquet. *Turn the Ship Around!* Penguin Publishing Group. Kindle Edition, p. 103.
5. www.youtube.com/watch?v=RcYjXbSJBN8
6. www.linkedin.com/pulse/scrum-very-confused-leadership-cliff-berg/
7. www.scrumguides.org/scrum-guide.html
8. L. David Marquet. *Turn the Ship Around!* Penguin Publishing Group. Kindle Edition, p. 97.
9. science.ksc.nasa.gov/shuttle/missions/51-1/docs/rogers-commission/Appendix-F.txt
10. www.exploratorium.edu/blogs/spectrum/two-cultures-physics-theorists-vs-experimentalists

6

What Effective Collaboration Looks Like

Agile reminded us that handing someone a document or a spec is not the most effective way to collaborate or share information.

Imagine if when you were about to first learn to play the piano, someone handed you a music textbook and said, "There you go—just read that. See you later!"

Not very effective.

Yet it became commonplace during the 1990s for large IT projects to define the work that way. A team would gather "requirements" from users and document them in a thick document—sometimes thousands of pages. The team was often contracted for only that phase—the requirements phase—and so they would then leave, and their documents were given to engineering teams, who were supposed to create a design that would satisfy the requirements. We're talking about any aspect of the design: either the functional design or the technical design.

That approach actually can work for a highly technical system, such as a communication protocol stack, but software that is used directly by humans is another matter: no programmer in their right mind chooses to implement a user-facing system purely from specs, but it was how countless projects were planned by IT procurement officers who had never written a line of code.

Even assuming that one could create a valid design based on a thick requirements document, the design phase was often done by another team that would leave, and the design document would be handed to a programming team to "implement."

To someone who is a building architect for homes or offices, that sounds perfectly reasonable, because any builder should be able to create a building from an architect's drawings. However, the reality is that in building construction, the architect is usually available to ask questions. In fact, they usually have a lot of say over construction decisions, because the architect knows about the physics of the building, and they are responsible for the safety of the design; and it is often the case that during construction there are decisions that need to be made, and the builder needs to know if the decision will compromise the intent of the design or the safety of the building.

Also, building design and construction is nothing like software design and programming, as we saw in Chapter 4. A building is three-dimensional, so you can see it and use your intuition about how things fit together; with software, you cannot. One can look at a building and see how close to done it is; with software, you cannot do that. When one builds a building, one often builds things that one has built before; but with software, one never creates the same thing twice.

Software is far more complicated than a building. A building is a hierarchical entity, with hundreds or thousands of different kinds of parts. In software, each line of code is a "part," and there can be millions of them, and they all interconnect in complicated nonhierarchical ways. You cannot visualize it.

This means that if you give someone a design for software, it is unlikely that the design will be enough for them to program from. Counterintuitively, the less detailed the design is, the better chance programmers have of being able to implement it, because they then do not have to understand the details but instead can invent their own details as they go. If they understand the requirements—what the software is supposed to do—they can make their own design decisions as they need to. The design then serves as a kind of guideline, and nothing more.

We have mentioned software several times, but these things are largely true for product development in general, because many products contain software. When electronics must be designed, the process has similarities to software, although there are important differences. Still, products today are often extremely complex, and so what we have been describing for software often applies.

That is not to say that designs are not important; they are *extremely* important. In fact, good-quality models and documents that describe how a system works are a good way to transfer knowledge between people quickly and effectively, as well as for establishing a shared concept for how a product should work. Designs are not sufficient, however, and should always be viewed as tentative, as changes might need to be made as issues are discovered during the development process.

Our point is that it is impractical to divide product development into completely separate phases. Product design is iterative. The require-

> **Agile 2 Principle:** Create documentation to share and deepen understanding.

ments need to evolve, because one cannot usually decide on all of the requirements until one has seen an early version of the product actually working, and a design needs to be fluid so that one can tweak the design in response to what one learns during testing. An initial design will rarely be correct.

It is also almost impossible for someone to understand a thick requirements document. It is like that music book. What people need is to *learn* the requirements. They need to have conversations about it. If you give them a requirements spec, they will not understand it until they have rewritten it in their own words and had many discussions about it.

The process of defining requirements, creating a design, and implementing the design is therefore an iterative and collaborative process.

Let's remind ourselves what the fundamental purpose of collaboration is. We collaborate when we have a shared interest or a shared objective. The goal of collaboration is to create a shared mental model. By collaborating, we synthesize our respective understanding to create a shared and consistent understanding. We also combine our ideas, which often leads to new insights.

Effective collaboration is essential in an Agile setting because Agile processes rely on rapid integration of every aspect of the work, rather than on long duration activities connected by handoffs. If collaboration is not effective, the Agile process will not be effective.

One of Agile 2's principles is that the whole team solves the whole problem. What this means is that everyone working on something needs to be "on the same page" about it. Silos of understanding

> **Agile 2 Principle:** The whole team solves the whole problem.

will not suffice: everyone needs to have some awareness and understanding of every aspect. Otherwise, synthesis is not possible. It is only through shared understanding—a shared mental model—that people can anticipate how changes affect each other. If one can anticipate how a change affects other aspects of a problem or product, including aspects that one is not dealing with but someone else is, one can alert those who might be affected. One might even be able to cover all of those other aspects oneself, avoiding the need to interrupt others.

Agility is achieved by reducing handoffs and transfers of information. It therefore might seem contradictory that collaboration, which partly entails a transfer of information, produces agility. However, what produces agility is *effective* collaboration, in lieu of handoffs and one-directional information transfers: working holistically instead of in sequence, addressing issues in real time instead of adding them to a lengthy queue, and thereby never waiting for anyone or anything. Waiting is a Lean "waste."

It is therefore essential to ensure that the real-time, collaborative processes being used are as effective as possible. Otherwise, one has not really gained anything by moving to Agile methods.

A Collaborative Approach

One could approach product development in a phased manner, where one starts defining requirements; then one (or others) starts creating a design; then others start implementing the design; and based on examination and testing of the implementation, one makes adjustments to the design and sometimes the requirements. In other words, use a waterfall process, but in an iterative and collaborative manner. Agilists do not do that, however, because if you do it that way, it takes a long time to get anything useful out; nothing is done until the whole thing is done.

Instead, Agilists prefer to work in an incremental manner, where one creates an overall vision of the product but then zeros in on a single feature of the functionality and then defines that well enough to design it at a high level; then they build a basic version of it and try it. This is often called a *minimum viable product* (MVP). If users like it, then the developers refine it or add another feature in the same manner. Gradually they build up the whole product and over time keep adding to it or improving it.

Such a process requires a great deal of collaboration, because each step is more of a sketch than a specification. For example, the requirements for a single feature are usually specified in the form of an Agile *story*, in which one merely states the objective of that feature—its desired outcome or what it enables a user to do. A development team then discusses the story with the person who wrote the story, until they feel that they understand it. They then design the feature at a high level, enough to start implementing it. In the process of implementing the feature, they often iterate on their design, which might require discussing the design with other teams or tech leads to decide on ways that the feature should integrate, in a technical sense, with other parts of the overall product.

Once the feature has been implemented, they show it to the person who wrote the story and get feedback. If the feature meets expectations, then the team focuses on a new story, but often the person who wrote the story will say, "Hmmm—now that I see it, I realize that it would be better if we changed this thing here." Thus, a new story is written. The team might start working on it right away, or the person writing the stories might say, "You know, that's not all that important—put it in the list of things to do, but at the end."

The interaction between the team(s) and the person writing the stories is therefore highly collaborative, as is the interaction between different product developers and among teams as well.

The approach of handing off documents between teams working sequentially is not workable for dynamic products. Documents are still useful; we do not want to discard the idea of writing things down. In fact, documenting product vision, requirements, and design are all really important, as long as one allows them to be refined and adjusted over time as collaboration progresses.

The Agile Manifesto advocated a shift of the emphasis from documents to collaboration, through its value of "Responding to change over following a plan" and its principles "Welcome changing requirements, even late in development" and "The most efficient and effective method of conveying information to and within a development team is face-to-face conversation."

Unfortunately, the manifesto gave no guidance for trade-offs pertaining to these things. As we have explained, the Agile community arose through Extreme Programming and other disruptive methodologies and already had a preference for extremes, and so it interpreted the manifesto's preferences as absolutes. Thus the preference for face-to-face conversation became a mantra. Agile coaches tended to always want every exchange to be verbal and in person. People used Agile as an excuse not to read emails, saying "Let's just meet and discuss it."

Not everyone collaborates well that way, however. Collaboration and conversation are not synonyms. Collaboration about complex issues often needs to occur over time, using a variety of communication formats, including writing, discussing, and diagramming. Also, an extremely important element of collaboration is thinking. When people collaborate, they exchange thoughts. However, one might need time to fully process something that someone has said. One's immediate verbal response might be the result of one's initial and shallow thought, but one might need deep thought to fully process it. Thinking is part of collaboration, and many people think best when alone and when given time to mull things over.

Respect How Others Work

Extreme Programming was arguably the spark that ignited the Agile movement. There had been rumblings about the dysfunction of how projects were being run, and people who understood software knew what was wrong. Agile methods were not new—they just were not collected under an umbrella called "Agile." People who had experience building software knew how to run successful software projects. Frederick Brooks' book *The Mythical Man Month* contained a great deal of "Agile" advice—25 years before the Agile movement.

We explained in Chapter 1 that the Agile movement quickly embraced a culture of extremes. One thing about extremes, and extremists, is that there is no choice: you have to follow, or you are censured. Those who challenged the extremes were called *Agile doubters* or other derisive terms.

Since the Agile Manifesto advocated for face-to-face communication, the community implemented that in an extreme way, and so any form of communication other than face-to-face verbal discussion was discouraged as "not Agile." Documents were initially frowned upon, until eventually people realized that they needed "Agile documents"— in other words, documents that were just enough and no more than what was needed. Whenever an Agile coach became aware of an issue that needed resolution, the knee-jerk reaction was always to gather the team to discuss it. The idea of having someone think it through and write down some ideas for circulation was rarely heard—it would have been viewed as an un-Agile approach or, worse, a waterfall approach.

The problem is, not everyone collaborates best through face-to-face conversation.

In particular, introverted people are known for preferring to think privately before discussing something. Introverted people also often prefer to write their ideas down rather than speak them. Writing requires less direct interaction, and verbal interaction with a group is taxing for an introvert.

People who are soft spoken—introverted or not—can struggle in a group. In a group discussion, people compete to get a sentence in. Aggressive talkers prepare themselves for the next opening and launch their sentence immediately after someone stops speaking. Soft spoken and thoughtful people do not communicate that way. When someone has finished speaking, a thoughtful person will still be mentally processing what was said, and they need pauses in the conversation to be able to have a chance. That is why they generally do better in one-on-one conversations but can never get a word in when in a group.

A Socratic leader who is facilitating the discussion will watch for these problems and proactively ask people what they think, giving the quiet ones a chance. It can still be challenging, though, because sometimes a subset of the group get "on a roll," firing ideas back and forth, and the leader might not want to interrupt that. Meanwhile, others who are less aggressive, and who might have great insights, sit quietly on the sidelines.

A facilitator also needs to protect those who have unpopular ideas. Sometimes unpopular ideas are actually the best ones, but it takes courage to voice an unpopular idea. That courage can come from a facilitator who makes sure that others do not mock the idea.

Group conversations are also not a good way to discuss things in depth. If there are many participants, they all want to be heard, so no one is allowed to have the floor for enough time to fully explain their thoughts. They get interrupted or someone interjects a tangential concern. An hour will go by, and yet only the surface of the issue has been scratched.

Writing is different. One can write down one's entire thought process, from end to end. Others can then read it, in its entirety. If people are unwilling to read what others write, it means that the group really cannot "go deep" on things.

We already mentioned that Jeff Bezos fought this "I don't read" trend by mandating reading and writing and says that "It was the best thing we did." Indeed, when we hear people say that they will not read long emails, our thoughts are, did they not learn to read and write in school? Did they not learn that those forms of communication are really important in society and business?

If someone has trouble reading and writing, one has to ask whether they are a good fit for job situations that require a lot of collaboration about complex issues. Perhaps their skills might be better in an operations role. Or perhaps they can improve their reading and writing. One of us had a writing mentor early in his career. The mentor, Ken, would force us to rewrite and rewrite, insisting that it be more concise and more to the point. If someone has trouble writing clearly and concisely, they need mentorship and coaching on how to do that, and they need to be told that it is important.

The Agile community's dogmatic insistence on face-to-face communication did not respect those who do not communicate well face-to-face, and it also "dumbs down" the kinds of collaboration and dialogue that we can have.

Agile 2 Principle: Foster diversity of communication and diversity of working style.

The insistence on open team rooms also created a great deal of stress for introverts, who need alone time. Today one often sees people in open spaces wearing headphones—not so that they can listen to music, but so that they can *tune out nearby conversations*.

All this pressure to be surrounded by a crowd all day is very abusive to people who do not want to live or work like that. It is like making a person with a knee problem stand on their feet all day. It is unfair and costly in terms of people's productivity and enthusiasm for their work.

Team Leads Need to Facilitate Effective Collaboration

Some teams have a team lead. Such a role is usually an "organizer" role. It might or might not have decision-making authority, and a team lead who is an organizer might also provide leadership in other ways. One of those ways, which strongly supports being an organizer, is to facilitate and encourage collaboration when collaboration is needed.

Too often team leads assume that others collaborate best the way that they (the lead) themselves do. For example, a team lead who prefers conversing instead of writing might presume that everyone else on a team also will communicate best if they all get together and talk.

Conversely, a team lead who likes to carefully write ideas down and trade emails might assume that others are so inclined and will be disappointed when some people do not read long emails or do not respond.

A member of a team should be sensitive to how others communicate best, so if it is clear that someone likes to send emails, others should try to make time to read them and to respond; similarly, someone who prefers written communication should notice when some others prefer to discuss things in person and attempt to have conversations with them. But people have limits, and members of a team are often busy. To be an effective organizer, a team lead needs to observe how exchanges among team members are progressing and take steps to improve things.

Team members also often do not notice when something needs to be discussed, because each person might be under time pressure to complete a task. A team lead should always be watching for issues that are going unaddressed. Indeed, each team member should as well, but it is a lot to ask—that someone focus deeply on completing a task, while also retaining focus on what everyone else is doing and what might be missed. That is why it is usually not a good idea to have a team lead also be heavily burdened with their own deliverables, because then they might not have the mental capacity to stay focused on how everything is fitting together.

A team lead should be always aware of what everyone else is working on, as well as how they are doing it. That puts the lead in the best position to notice when something is being missed.

One practice that can be effective is to have occasional team-wide discussions about how things are fitting together. Doing that is particularly useful when the work is in a phase that has lots of unresolved complexity. This is often the case in the "middle" of a project, major development increment, or the development of a very complex feature set. At the beginning of such an initiative, there need to be discussions to settle on a design; but as development progresses, issues arise that were not apparent at the beginning, and the issues can often be multifaceted and involve aspects that many different people are working on. Keeping everyone apprised of the approach that everyone else is taking is very effective for revealing additional issues so that they can be discussed and resolved.

A team lead should be listening for complaints that team members are struggling or confused about how things fit together and encourage discussion

Agile 2 Principle: From time to time, reflect, and then enact change.

with others who might be affected. People will naturally want to discuss things in the way that is best for them, but a team lead should be thinking about what process might work best for all involved.

For example, if someone is concerned that they are using data structures (in a database or in a spreadsheet) and others are frequently changing the structure and not telling others, not documenting the changes, or not updating other parts of the product that might be affected, then a quick meeting to brainstorm on solutions to the issue might be the best approach.

On the other hand, if there is a complex issue, such as how changes should be integrated spanning multiple parts of a product and several people have voiced conflicting solutions to the problem, the team lead might ask those individuals if they would each write up their thoughts using a short format—bulleted, or what Jim VandeHei, founder of news startup Axios, calls "Smart Brevity." The writing style is described by marketing strategist Lucas Quagliata as "short and to the point (they often literally use bullet points), just the right amount of context, and often a conclusion of sorts ('what it means' or 'why you should care')".[1]

These short solution descriptions can be distributed to the team, and people comment. After a few exchanges, the pros and cons of each approach will have been revealed, and it might be a good time to have a meeting to discuss them and decide on an approach.

If someone feels they are not good at writing, even a short description, the team lead could ask someone who is good at writing to help that person. Have the person who prefers talking explain their approach and the other write it down.

The goal is to ensure that collaboration happens, that important issues are not overlooked, that collaboration is effective and tailored to the level of complexity of the issue, and that resolution is achieved in a timely manner—that is, that the discussion does not become open-ended and never lead to a decision. A team lead is a kind of organizer and orchestrator to make sure that this happens; because while everyone on a team is responsible, a team lead can be a huge help.

Every Interruption Is Costly

Today's "always-on" culture is harmful for one's ability to focus. It can translate into continual interruptions. The "Agile team room" is also a distracting place and therefore not conducive to focus. Focus needs to be balanced with collaboration, but the team room and always-on "push" communication tools such as Slack and Teams can make concentration difficult.

A team lead acting as an organizer should hesitate before asking for a tool or a team seating arrangement that forces the team members into a single mode of work—collaborating or focusing—rather than allowing them to consider how best to balance between these two needs. Instead of dictating, one might ask the individuals on the team and remember that they might not all have the same needs.

Agile 2 Principle: Make it easy for people to engage in uninterrupted, focused work.

On the other hand, people often do not know how they will thrive. This brings to mind the endless debate about "basic income," whereby everyone is given a monthly allowance by their government. The assumption is that people will then pursue what is fulfilling to them. Some people will do well under such a system, while others will dissipate. Some people need to be told they must work so that they do not play computer games all day or some other empty pursuit, while for others work interferes with their ability to create and develop to their potential.

Agile 2 Principle: Individuals matter just as the team matters.

If one asks people what seating arrangement they will enjoy the most, we suspect that most will want to sit with others, in an open arrangement. The reason is partly that they are used to it and see it as normal (it was not always that way), but also that it is *more fun!*

Being fun, however, is not the same as being productive.

Teams should reflect on how they are most productive and effective, in terms of tools, seating, and other factors. Management should not impose a single model on all teams. Individuals on teams also have different needs, and not all members of a team should be expected to work the same way.

Some people might work better remotely. Others might need frequent interaction with others. People vary greatly in that regard.

One thing that is universal, though, is that interruptions are costly, because they shatter mental flow and focus. Even people who benefit from frequent collaboration might need to focus from time to time, without interruption. Therefore, tools and arrangements that constrain people to the mode of always on or always off should therefore be reconsidered. Isolating everyone as a practice is as bad as putting them all together.

Push tools should be avoided for work that requires a lot of focus. It is appropriate for operational activities, where real-time communication with lots of individuals and groups is central to the work, but push tools are detrimental to work that requires concentration. Fortunately, most push tools can be configured so that they can be used in a "pull" mode.

Standing Meetings Are Costly

The Agile "standup" meeting is a mainstay of Agile teams, and we explained standups and some problems with the practice in a prior chapter.

Standing meetings in general are problematic. They are scheduled under the assumption that they are for collaboration and therefore "are Agile," but standing meetings are really a relic of old-school work practices that are schedule-based instead of event-based.

When someone describes an "impediment" during an Agile standup, what we want to know is, "Why did you wait until the meeting to speak up? Why did you not raise the issue when it arose?"

If people—say, a group of development managers and team leads—schedule a weekly meeting for receiving an update, what we want to know is, "Why do you need us to give you an update in person? Are our dashboards insufficient? If so, tell us how, and we will fix them."

If the response to that is, "But we want to hear from you what you think," then our response to that is, "Yes, that is important, but does your need to know what we think arise just before the weekly meeting?

Or was there a point during the week when you actually needed to know? And if it was the latter, why did you not send an email, and ask for a ten minute chat?"

The response to that might be, "But all the managers need to have a chance to discuss what you share," then our response to that would be, "Why not first wait to see if there is actually something to discuss, and if so, then email those managers to set up a quick call?"

The response to that is surely, "But their schedules are all full—they would not have time for a call that we could fit into all of their calendars."

And there is where the blocker usually is: decision-makers usually have full calendars, but their calendars are full of standing meetings!

So this is a catch-22.

The solution is to get rid of most standing meetings and start managing by dashboard, by event, and by inquiry.

Agile 2 Principle: Respect cognitive flow.

The full calendar that so many managers have is a huge dysfunction, and it is a major blocker for an Agile organization. The culprit is standing meetings.

A meeting should never be scheduled to inform; information can be placed in a dashboard or a shared folder or collaboration tool or sent by email. Meetings should be for *discussion*. Gathering a group for a meeting is costly, in terms of people's time and focus, and it should be reserved for when it is already realized that discussion is needed.

The cost of unnecessary meetings is high. According to organizational psychologist Dr. Marta Wilson,

"Meetings consume much of the average professional worker's time. Too often, they paralyze an organization, sapping the time and energy of all concerned. . .Meetings are not needed to discuss trivial subjects or to disseminate information that could be distributed using other means."[2]

Some meetings are about maintaining a relationship. These tend to be one-on-one meetings. That is a different kind of meeting. Its *main* purpose is not to share information or to discuss an issue; it is to build or maintain trust and a personal link. For that kind of meeting, one would still meet even if one had no new information to share and nothing to discuss. The relationship is what that kind of meeting is really about.

One-on-one meetings are extremely effective for "going deep"—that is, discussing something that enables two people to unravel an issue and reach a shared understanding. In a one-on-one meeting, others are less likely to interrupt or drive the discussion down a tangential issue. Since a one-on-one discussion occurs by taking turns, one's turn is always next, and so one can bring the topic back to the desired focus. True, one party can dominate, but it is far less likely than in a group. That is why one-on-one meetings are useful for helping someone to understand an issue well enough to be able to act on it. If you need to persuade someone, set up a one-on-one meeting.

But do not use one-on-one meetings for sharing information. That wastes people's time. Use them for discussion or for maintaining a relationship.

If people eliminate their standing meetings, they will quickly find that they are out-of-date and "out of the loop." This is because they have come to depend on standing meetings. To do without standing meetings, one must change how one communicates and stays current. One must create real-time information radiators and get others to keep them up-to-date. One must become issue-focused and reach out to others to check one's understanding, instead of waiting for a meeting to present the opportunity to discuss the issue. One must also think about who is needed for resolving each issue. Instead of bringing together the entire usual group, reach out to those who you believe are affected and merely notify the others in case they are interested and want to join you.

Deep Exchanges Are Needed

If people's calendars became more open, they could have longer meetings and be able to "go deep." Instead of a manager saying "I have a hard stop in half an hour, so just tell me your ask," the manager can say, "Explain to me what is really going on here, and connect all the dots, please."

That will lead to a much more intelligent organization that makes much better decisions.

Organizations are becoming "dumbed down" by the inability to have deep conversations. Business and technology are far more complex than they were 10—let alone 20—years ago, and yet managers make less time to try to understand things.

They believe that they are needed in all those meetings, but the truth is that a lot of the standing meetings that managers attend could be replaced by emails, dashboard, or direct interactions only as needed.

By not engaging in deep conversations, managers operate with only superficial understanding of the many complex issues that they face. They want to know how something

Agile 2 Principle: Foster deep exchanges.

can be summed up in one sentence—"give me the gist." But in obtaining only "the gist," one does not actually understand: a "gist" is someone else's conclusion.

To truly understand something, one has to explore the phenomena that are occurring and their cause-and-effect relationships. To have discussions that lead to understanding, one must take an inquisitive approach: "Tell me what is happening" followed by "Why is that?" and then another "Why is *that*"? During such a conversation one will learn things that contradict one's own understanding, and that should lead to "But I thought. . . ." The dialogue should reveal core assumptions—things that can be tested or validated by talking to others or looking at some data.

If a manager—any leader actually—takes an inquisitive approach, their goal is not to make a decision; rather, their goal is to *understand*. In that moment, they become a Socratic leader, driving toward shared understanding.

John von Neumann, the famous physicist and mathematician, was known for leading thoughtful discussions that led to understanding. In Walter Isaacson's book *The Innovators*, he describes how von Neumann operated: "Pacing in front of a blackboard and ringleading the discussion with the engagement of a Socratic moderator, he absorbed ideas, refined them, and then wrote them on the board. 'He would stand in front of the room like a professor, consulting with us.'"[3]

An important element of such a discussion is that the best idea should win. "Von Neumann was open but intellectually intimidating. . . But Jennings sometimes [pushed back]. One day she disputed one of his points. . . But von Neumann paused, tilted his head, and then accepted her point of view. Von Neumann could listen well. . . ."

Exchanges like this should not be fast-paced. People need to feel like they are not wasting your time or that you are in a rush. They need to feel like the person with authority in the room is really trying to understand things. That is the only way that people will risk sharing their full understanding. People are leery of sharing their true thoughts with a superior; it makes them vulnerable, because they could be wrong, or the boss could disagree. If a boss disagrees, it might lower one's standing in the boss's eyes. That is why it is essential that a manager demonstrate that understanding is their goal, that they value it when people are able to have honest and thoughtful discussions and are not just looking for "the answer."

The cost of not having deep discussions is high. Managers are strategic decision-makers, and they can make strategic mistakes. A big mistake that we have seen IT managers make when adopting Agile methods is to eliminate all official testing roles. The decision is usually made in an effort to "be more Agile" since some Agile articles claim that Agile teams perform their own testing. However, that is a terrible oversimplification of the situation, and that oversimplification can lead to a simpleminded decision, such as eliminating all testing roles—which is an extremely costly decision.

To apply Agile ideas, one must understand them. Much of Agile has been oversimplified, leading to terrible implementations. The elimination of testers is but one example. To make good decisions, one must understand an issue, and Agile issues are anything but simple.

That is why deep conversations are needed.

The Whole Remote vs. In-Person Thing

Reed Hastings, the founder and CEO of Netflix, describes remote work as "a pure negative."[4] He says that "Debating ideas is harder now."[5] When asked the question whether he sees any benefits of remote work, his response was "No. I don't see any positives," so his view is pretty extreme.

The article was posted to Slashdot (slashdot.org), and one commenter wrote, "Sounds like an extrovert who misses being able to randomly walk in on people and jump into conversations. I've absolutely LOVED not having to sit in a office and deal with other peoples incessant chit-chat while trying to get my shit done."[6]

Microsoft conducted a study of the use of its Teams collaboration tool during the COVID-19 lockdown. It found that "while Microsoft salespeople have significantly increased their collaboration time with customers, people in our manufacturing group have focused on streamlining and optimizing connection points with a growing number of supplier contacts."[7]

Unsurprisingly, Microsoft also found that online meetings increased but that the meetings became shorter: "weekly meeting time increased by 10% overall—we could no longer catch up in hallways or by the coffee machine, so we were scheduling more connections—individual meetings actually shrank in duration."

One might think that if people are at home, out of sight, they might not pay as much attention during meetings, but Microsoft found that "multitasking during meetings didn't spike even though people weren't in the same room."

People's online connectivity also increased: "We also measured networks across more than 90,000 Microsoft employees in the United States. . .we expected to see them shrink significantly, given the rapid shifts in environment, daytime rhythms, and personal responsibilities. Instead, we discovered that most employees maintained their existing connections. Even more encouraging, most people's network size increased."

Some people have trouble working at home: the lack of structure and distractions within the home can present serious challenges. However, it is important to realize that being at home versus being in an office affects different people differently.

Some people focus better at home. For example, a 30-year-old graphic animator interviewed by BBC News reported that being at home made it easier for her to focus, compared to being in an open office space with people around.[8] She also said, "If I was having difficulty concentrating or finding inspiration in the office, I just had to push through that feeling. Now I can go and have a chat to my husband or play with the cat or sit in the garden for 10 minutes and come back fresh."

Some people have also found that being home has improved their mental health as well as their bonds with their family. The BBC article described the impact on Nirali Amin, a bookkeeper: "Nirali believes the bond she has with her children is stronger than ever before. Prior to lockdown, she feels they had less time to discuss family matters in depth together."

We are not saying that events that precipitated the 2020 exodus from offices were a good thing. Indeed, many people were adversely impacted. Rather, we are pointing out that among those who shifted their work to a home office, some have benefited and some have not.

In other words, it depends. Again, one size does not fit all. People differ—a lot.

There is no optimal solution to working in the office or working at home. What will be beneficial for some people will be anathema for others.

Agile 2 Principle: Foster diversity of communication and diversity of working style.

Perhaps organizations will have to decide on an office-by-office basis. An organization might have to decide if a group will be on-site or off-site. Such a decision might weigh the kinds of collaboration that need to occur, as well as the personality types that will be attracted to that work.

The company Zappos, which pioneered a radical organization structure known as *holacracy*, famously let go of a large percent of its employees, on the grounds that they did not fit the new culture that Zappos wanted to create. In other words, Zappos realized that for holacracy to work, it needed a certain personality profile.

Organizations might have to engineer their culture in this way by identifying the skills and experience needed and the personality types needed to create a success in the chosen domain according to the chosen strategies. Those considerations might then dictate whether the organization should create an on-site workforce or an off-site workforce.

On the other hand, there is a lot of research on how important diversity is for an organization's success, and when an organization screens out people who "don't fit," they are actually decreasing diversity, perhaps in unintended ways.

How to Make Remote Work *Work*

Even on-site workers need sometimes to work remotely. Also, an organization might find that its people have a mix of personality profiles, and it is not feasible or fair to do what Zappos did—that is, let go of all those who did not "fit."

That means it might be necessary to enable people to be on-site or remote or perhaps have a hybrid consisting of some on-site days and some remote days. On any given day, some people might be on-site and others remote. In such a situation, finding ways to make that work is imperative. The question is then, what are the best practices for remote work and for mixed on-site and remote people?

There are books about how to support remote work. An excellent one is Johanna Rothman's *From Chaos to Successful Distributed Agile Teams: Collaborate to Deliver*. So we will not dive into remote work practices here. What we will say is that to make remote teams effective, a different approach is needed. You have to *make* it work. Intentional practices to support remote workers, regardless of their percentage of your workforce, are the key.

You will need to take steps to make sure that discussions and decisions are transparent, because they can too easily occur in private email exchanges or direct messages. Do not replicate the "men's room" decision-making of yesterday, in which two or three men have a private conversation in a men's room and make a decision, with no one else's input.

Also, the Agile default of "get everyone in a room and talk" is not always the most effective when in person, and it can be just as ineffective remotely if there are multiple time zones involved. Ironically, we feel that remote meetings can actually empower those who would stay silent in an in-person meeting, because voices and physical presence tend to all become equalized. We already mentioned in an earlier chapter that the leaders who tend to emerge for remote work are the organizers, rather than the ones with the "most personality."

Introverts benefit most from remote work. Andy Luse, a management scientist at the Spears School of Business at Oklahoma State University, measured the personality traits and thinking style of more than 150 business students and then assessed their preferences for virtual teamwork. According to Luse,

> *"[Extroverted] individuals would much rather work face-to-face as compared to virtually, which will lessen the energy they get from the interaction. . . . Conversely, introverts expend energy with social interaction, so while they are more apt to work alone well, they are also better at adapting to a virtual environment given it involves less face-to-face interaction and is thereby less taxing mentally."[9]*

Just as being in an office might favor those who are extroverted and socially aggressive, being remote might favor those who write. It is important to make sure that everyone has a voice and that all ideas get heard and are considered.

Do not make the mistake that was made by the Agile movement, of assuming that everyone works the same way. Some people communicate best verbally, and others communicate best through writing. Agile 2 has a principle, 6.2, which reads, "Foster diversity of communication and diversity of working style." The intention of that is to make sure that if some people contribute to a discussion during live chat sessions and others send emails or write comments in an online document, make sure that both get vetted and both contribute to advancing the issue. Do not let one form of communication dominate, unless it is clear that *everyone* prefers that form.

Preferences for collaborative work vary by person as well. Some people will favor real-time group-based activities, including artifact creation; they will want to brainstorm and create the content as a group. Others will not be able to work that way at all and will want to share ideas in real time but then create an artifact draft on their own and request review—in other words, a back-and-forth handoff.

If people are in different time zones, a handoff might be the only practical approach. Someone needs to act as a facilitator for this to make sure that the process is working well and that everyone is contributing in the way that they are most effective.

Talk to your collaboration partner(s) and decide how you will work together. Then try it, reassess, and pivot. But make sure that no one is being left out or left at a disadvantage.

If people need to work in pairs, pairing people who work in the same way is really important. Allow people to decide who they pair with (some people are a horrible pair because they work so differently). As we explained, some people are on-the-fly thinkers who act as they think, and others need to think fully before they act. If these two different types are paired, they will not be able to work together unless one of them compromises and works in a way that is not natural for them and therefore not optimal.

Online information radiators, with unobtrusive "pull" notifications of important updates, are also essential to share information from *all levels* of the organization. If there is nowhere to look to find information, people have to ask, or they will simply go without knowing. Information radiators provide one-directional communication that should never have to be bidirectional. Discussions—verbal or written—should be reserved for deep exchanges, meaning conversations that try to get to the heart of an issue. Merely obtaining information should be something that is easy and does not disturb anyone.

The most challenging situation is when some people are in an office and some are remote. In that situation, a real-time meeting is awkward. Language matters. Accents, dialects, and speaking clearly and loudly are essential for calls in which some people are in a room together and others are remote—often, the people in the room are not heard easily by the remote people. A facilitator should assess if people have trouble hearing and get the soft speakers to speak up or alternatively repeat what they said in a way that is easy to hear. Some people need speech coaching! It is possible to train people to articulate loudly and clearly, as if on a stage.

Are Remote Teams a Trend?

In 1997 one of us wrote to the COO of Barnes & Noble to warn him that in 10 years people would be reading books on computers and that print books—and therefore bookstores—were at risk.

He did not respond.

When we told friends and colleagues this, the reaction was almost always, "No one will want to read a book on a computer." They imagined the bulky cathode ray tube monitors of the day, with low resolution, and could not visualize what computers might be like in 10 years or that they would ever want to read a book on one.

Ten years later the Kindle came out, and three years later the first iPad. Print books are still widely used, but people now routinely read books on "computers"—that is, their phones, tablets, or e-readers—and bookstores have become a rare thing. It happened a lot faster than people imagined it would. These kinds of trends can be like a cascade: change seems far off until suddenly there is a tipping point, and then change comes rapidly.

The same thing is happening with remote work. Alvin Toffler among others predicted long ago that the cost of electronic communication would drop so low that people would stop commuting. Commuting arose during the 20th century, when work shifted from the farm to the factory. For all of human history people did not commute to an office—until the 20th century. Now in the 21st century, we are seeing the end of the office. Commuting to a faraway office was a blip—an aberration.

Wall-sized displays are coming, applied as wallpaper. Why travel 10 miles to an office or to meet someone for business when you can see them lifesize and crystal clear on your wall? To see family, sure: there is nothing like being there in person to give them a hug. But routine business? People will just click an address book entry and initiate a video call—perhaps it will even be 3D.

That is the inexorable trend, and there is no stopping it, short of the collapse of industrial society.

There are powerful economic forces driving it as well. Being able to collaborate remotely means that one can include people who are anywhere. One can now have high-resolution low-latency video calls between the United States and Australia or between Australia and India. That means one can find the best people for a job, instead of being geographically limited; and people have a whole world of opportunity, instead of being constrained to within 20 miles of their home.

Remote work is therefore an unstoppable trend. Even those who prefer to show up in person will find themselves increasingly displaced by those who are able to call in remotely but perhaps are the perfect fit for the job, even though they live far away. Distance will not matter.

Remote is the future. Whether it happens now or in five years or ten, it will happen.

Notes

1. medium.com/that-good-you-need/smart-brevity-7ff4d336a989
2. Dr. Marta Wilson. *Leaders in Motion: Winning the Race for Organizational Health, Wealth, and Creative Power.* Greenleaf Book Group Press. Kindle Edition, p. 127.
3. Walter Isaacson. *The Innovators.* 2014, p. 107.
4. www.wsj.com/articles/netflixs-reed-hastings-deems-remote-work-a-pure-negative-11599487219
5. tech.slashdot.org/story/20/09/07/156202/netflixs-reed-hastings-deems-remote-work-a-pure-negative
6. tech.slashdot.org/comments.pl?sid=17134144&cid=60482122

7. insights.office.com/workplace-analytics/microsoft-analyzed-data-on-its-newly-remote-workforce/

8. www.bbc.com/news/uk-53580656

9. www.bbc.com/worklife/article/20200601-the-personalities-that-benefit-most-from-remote-work

7

It's All About the Product

Gene Kim is widely viewed as a spokesperson for DevOps. (DevOps is a set of ideas pertaining to delivering software rapidly at scale.) His 2013 book *The Phoenix Project* catapulted DevOps into the awareness of the wider IT community. He has also defined what he believes are the core principles of DevOps. He calls these the "Three Ways [of DevOps]".[1] On the first "way" he writes,

The First Way emphasizes the performance of the entire system, as opposed to the performance of a specific silo of work or department — this can be as large as a division (e.g., Development or IT Operations) or as small as an individual contributor (e.g., a developer, system administrator).

His point is that what matters is the product and its platform, which he refers to as the *entire system*. These matter more than the work unit, meaning any organizational unit such as a team.

While DevOps methods arose in the arena of software, the ideas are entirely applicable to all kinds of products. This is especially true as more and more products are created through digital design and manufacturing methods.

There is a strong focus on teams in Agile thinking, as well as in DevOps thinking. This is because work is more scalable if teams are able to operate independently. However, for a multiteam product, no team is completely autonomous. Otherwise, there is not a product. Instead, there are many products—one for each team.

The product needs to be the focus, not the team. When one creates a design, the design is for the product, not the team. When one creates a test strategy, the strategy should be for the product, not a team.

A team is a group of humans, and a team needs to decide how to work together. However, its methods should be subordinate to decisions that have been made for the product. If the team cannot accommodate some product-level methods, it should bring that issue to the other teams and resolve it. Otherwise, other teams might expect things to be done a way that they are not being done, and confusion or inconsistencies will result.

There are four primary aspects of issues and decisions that commonly arise in the course of product development. These were presented in Chapter 5 in the section "Common Types of Product Leadership That Are Needed."

- **Product value**—the product vision and what the product does, including its UX design and its functionality
- **Product technical design**—how the product works, internally
- **Product development workflow**—how the teams work together, and with other stakeholders, and the workflow and human dimensions
- **Product development technical practices**—the technical aspects of the workflow

Each of these aspects needs to be designed.

The first of these aspects is traditionally thought of as "product requirements," but successful product companies know that viewpoint is inadequate today. Today's products often need to be extremely engaging and creatively conceived. This is why product design is so important today—it pertains to these things, and we will address Agile 2's ideas pertaining to product design in the next chapter.

Product *technical* design is traditionally thought of as driven by the product requirements. However, today's products are often driven by new technologies that make new things possible. For example, the availability of new methods for creating small metal parts on demand, such as metal laser sintering, has driven down the cost of small jet engines and made them more feasible for many applications.

Indeed, the dropping cost and automation in the just-in-time fabrication of materials is behind today's revolution in personal aerospace, such as the turbine-powered products being developed by Zapata, which are designed using 3D solid modeling.[2] And of course the big example of new technology making new things possible is the Internet, which made Amazon possible.

The flow of ideas between business thinkers and technology thinkers is therefore not one-directional; it is bidirectional. A product vision can originate from someone who understands a market or from someone who understands a technology. The best products often come from someone who has a deep understanding of both.

No matter where a product idea originates, its design is not a matter of mere execution. Performing the technical design of a product is challenging and requires inspiration and great skill. The technical design of a product requires vision every bit as much as the functional design does. And just as the functional vision must be ever-present among those who are executing the vision, the technical vision must be ever-present among the execution teams as well.

A vision guides people in their decisions. Implementing a product is a series of decisions. Every line that a programmer writes is a decision; every circuit or part that an engineer designs is a decision. If the vision is not present in the minds of those who are executing the vision, the resulting product can turn out to be a jumble of parts that meet the specs but that are not true to the vision, desirable to users, or valuable to the organization—a kind of Frankenstein's monster.

Product development has a human aspect and a technical practice aspect. Business stakeholders assume that development teams know their job and that they will come up with a delivery process that suits them. That is akin to saying, "Businesspeople know the market and will come up with a product concept that suits them," in effect dismissing the complexities and the risk of getting it wrong and the need for a thoughtful process for how the product concept should be developed and validated.

The product development process is itself a system that needs to be thought through and designed, and its design needs to be continuously refined over time.

Agile 2 Principle: Business leaders must understand how products and services are built and delivered.

Today a product development process typically consists of a great deal of tools and automation, and that process is itself a product. It is the product that builds and validates the market-facing product. As any product, the delivery process needs to be designed and refined, and it needs to be driven by a vision, which today is usually informed by Agile and DevOps ideas.

The product development system is sometimes the thing of real value. Recall our race car team example. As another example, Amazon's "product" is its online catalog of products and the systems it has for obtaining and distributing those products. However, Amazon is not valuable because of that catalog or its distribution system; it is valuable because it has a *system for maintaining* that catalog and distribution system, which are ever-changing because they deal with a huge number of vendors, and those vendors are constantly changing their products. Amazon needs to keep pace with that and be able to optimally monetize its arrangements, which requires rapidly updating its catalog, deals, and other features.

If they could not respond rapidly, they would not be able to keep up with the market, and a competitor would step

Agile 2 Principle: Technology delivery leadership must understand technology delivery.

in. The value is in the speed and breadth of their platform's ability to respond to change. In other words, it is the agility of their platform that provides the real value, specifically, the ability to make changes and push those out right away, with confidence.

The business value rests on their agility.

This is important, because it means that *how* they deliver their services is as important or even more important than the services that they deliver. The IT

Agile 2 Principle: Technical agility and business agility are inseparable: one cannot understand one without also understanding the other.

functions that provide their technical agility are central to their business agility and to their market value.

What to Prioritize

There are complicated formulas for product feature prioritization. One is known as *weighted shortest job first* (WSJF). The basic idea of WSJF is that "low-hanging fruit"—things that are "short" in terms of the time to complete them—should be given priority, and things that are high value should also be given priority. Balancing these is the intention of the WSJF formula.

One can also factor in the cost of delay—that is, the opportunity cost of delaying a market-facing feature. If one wants to get really sophisticated, one can use a "real options" formula to compute the expected net present value of each product feature choice and choose the one with the highest expected value. One of us wrote an entire book on that subject once, but he has come to believe that people usually do not have the data to accurately make such calculations, and even if they did, it is not worth the effort, because human gut feeling is often better, and it is a lot faster.

One important thing that WSJF leaves out is the risk that a feature presents. Some features are much more difficult to implement than others. It is not just that they take longer; it is that they have unknowns, and so the time to complete the feature is much harder to estimate—in fact, it might turn out that the feature cannot be done at all.

Things that are complicated take longer, but they are also the things that are most likely to be a source of problems later—problems that need fixing. These can arise during later stages of testing or even during use by users.

Uncertainty is costly because it puts downstream things at risk, and risk is an expected cost. If a future feature is believed to have a large value to users, but the feature relies on a precursor feature that is difficult to implement, then the future valuable feature is at risk until you succeed in implementing the precursor feature.

For this reason, it is often sensible, and economically optimal, to implement the risky things that are

Agile 2 Principle: Integrate early and often.

precursors to the most valuable things. The Agile 2 principle "Integrate early and often" pertains to this. Its advice is to try to identify downstream issues on a frequent basis and not postpone steps that might reveal problems later. In other words, do what you can to reveal unavoidable problems as soon as possible. Another way to say this is to "shift risk left"—that is, push it to the left of your current timeline, assuming that your timeline goes from left to right (a Western cultural norm).

This is not a project management book, so we are not going to give you a method for prioritizing features. The point we want to make is that feature prioritization should be done with the product in mind—the whole product. And it should consider the things that might go wrong and be informed by a thoughtful consideration of technical (as well as business) risks.

A Product Should Be Self-Measuring ▬▬

You need to stay aware of how well the product is performing when people use it. If you have a car, you want to know if your car is running well, right? That is why your car has gauges and warning lights. It is the same thing.

That is why a product should have built-in feedback loops. These should be designed at three levels.

- **The infrastructure level**—how well are the underlying platforms running? Are the networks available? Are message queues saturated?
- **The product engineering level**—how well are the deployed parts of the product working? Is everything up and running and running well, in a technical sense?
- **The business usage and outcome level**—which features of the product are people using? What level of performance do they receive? And are the desired business outcomes being attained?

Levels 1 and 2 are almost always covered to some degree, because engineers focus on those. Level 2—the product engineering level—is sometimes not covered *well*, though. To provide effective feedback loops at the product engineering level, developers (or operations personnel) must insert logging and monitoring at various points within a product. It is easy to go overboard so that so many messages are generated that no one pays attention, and then when there is a real problem, it gets missed. Also, the data collected must be correlated so that if a problem shows up deep within an application, it can be matched with what a user was doing.

The really challenging level is level 3. It is challenging because it must be defined by the business stakeholders.

Agile 2 Principle: The only proof of value is a business outcome.

They need to identify which parts of a product they want to monitor, and in what way; and they must identify how a business outcome can be measured or estimated based on how a product is being used. For example, has a new feature translated into more purchases over time? The cause and effect might be somewhat indirect, so measuring it might require collecting and labeling data and then performing business intelligence analysis on the data afterward. That cause-and-effect measurement is a feedback loop. If a feature translates into better outcomes, the feedback is positive; otherwise, it is inconsequential or even negative.

Development System as Product

Since the product development system is itself a product, it *also* needs feedback loops. These feedback loops do not measure how well the actual product is working; they measure how well the product development system is working. (Remember, when we say "development", we are including delivery of the feature to production usage.)

In other words, they measure how effectively and rapidly teams turn ideas into solutions that can be tried by users, as well as the quality of those solutions, within the parameters of the solution idea.

For example, consider a banking application for which the business and delivery teams realize that users probably need a new feature to enable the user to request an increase in their credit limit. How fast can that feature be developed? And once developed, how robust is it? If a feature is riddled with problems and causes production incidents, it is not useful, and so measuring speed of development is not a good idea unless one *also* measures quality.

The development time is known as *lead time*, or *cycle time*, depending on whether you include the time from when a business stakeholder requests a new feature be built to when the teams actually find time to start working on it.

It turns out that the time to develop a new feature is mostly consumed by the time required to test it. Most of the work in product development is spent checking if something works and, if not, figuring out why and then checking again. Programmers, for example, spend an amount of time coding a new feature and a comparable amount of time figuring out why their code does not work. Our own honest estimate is that upward of half of a programmer's time is spent in the state of "It should work, but it doesn't—why not?!"

Teams often measure their *velocity*—the rate at which they complete Agile stories. An important fact about velocity is that it is a team metric; it is not tied to the product. Different teams will have different velocities, and you cannot compare them, because the story velocity value is not pegged to anything tangible. It is based on story point estimates, which are purely a gut feeling value and have no actual meaning. But a given group of people tend to assign consistent story point values. That consistency is what makes the velocity metric useful for projecting how much work *that* a team might get done in a future period of time.

The primary focus needs to be on the *product*, not the team. If a team can run product-level tests, that is ideal; but often teams cannot act on their own across the entire product, especially for complex products. It is important to not lose sight of the fact that the product is what matters.

We mentioned that much of a product developer's time is spent figuring out why things do not work. Therefore, if one adds a new feature to a product, the major task is not to design and implement the new feature, but to get it to work. The only way to know whether it works is to test it. One can test a feature in isolation or test it as part of the *whole product*. The latter is referred to as *integration testing*, and if truly the whole product is included in the test, then the test is referred to as an *end-to-end integration* test. Managing the thoroughness of these tests is the primary way to manage risk for continuous automated processes. We will explain testing flow and real-time risk management more in Chapter 9.

However well you build something, the final test of quality is what people think of it when they see it and

Agile 2 Principle: Integrate early and often.

whether it holds up under use. For example, is it buggy? Is it slow? Do the server applications experience incidents?

Any problems resulting from use of the product should be categorized according to root cause and provided to the teams so that the teams can identify (a) which aspects of design and implementation might need to be improved and (b) which areas of testing might need more thorough coverage. These are critical feedback loops for quality. This is not team feedback; it is *product* feedback.

People view product usage problems or "incidents" in one of two ways: a problem indicates a design flaw or a problem indicates a testing gap. In reality, it is both. Any field usage or production problem can be seen as something that was not thought of during implementation, but it can also be thought of as something that "escaped" the test suite. If one does not address the gap on both fronts, then one has an incomplete process. Deciding how to balance effort in these two areas is a judgment call.

We can now start to assemble metrics and practices that provide the feedback that we need on the delivery process.

- Feature test thoroughness
- Team velocity
- Feature lead time and feature cycle time
- Product incident rate; user complaints

The first one, test thoroughness, is a "leading" metric, and it pertains to the behavior of the development team members. It is a behavioral metric. Behavioral metrics are useful for encouraging (through measurement and awareness) behaviors that one believes will result in better outcomes. The second, third, and fourth of these are "trailing" metrics; that is, they tell you how well you are doing.

One can be wrong, however, about which behaviors will help, so behavioral metrics should always be

Agile 2 Principle: From time to time, reflect, and then enact change.

viewed with some tentativeness; it is often the case that someone uses atypical behavior but is still able to achieve good results. For example, a programming team might do little testing yet produce highly reliable software. That might be the case if they use an alternative method to ensure quality, such as rigorous code review. There are also teams that swear by mob programming, although others abhor the approach. Teams and leaders should therefore frequently reflect on the behaviors that are being encouraged and measured.

Behavioral metrics should be behavioral nudges but rarely should they be used to police or enforce a way of working. An exception is if they pertain to compliance rules that are beyond the organization's control or are there to protect against a severe risk; for example, measuring whether people wear hard hats in a construction zone informs about safety and is not a rule that should be considered to be flexible.

Generally one should not measure or score people on *how* they do things: behavioral metrics should be there to encourage, but not punish, and the outcome should be what matters in most cases. It should be pointed out, however, that outcomes are not all that matter. Behavior matters too. For example, if someone is successful by being a tyrant, the organization might not want to allow that behavior to be rewarded.

However, one should be flexible with respect to behaviors that are not deemed to be dysfunctional or against the culture that the organization is trying to maintain.

There are leading and trailing metrics that make sense to collect for any kind of product or service creation system. Whoever gets work organized for the teams should be identifying those metrics, with input from those doing the work as well as leadership, and should be arranging for them to be measured and displayed in a visible way, with updates as close to real time as possible. Everyone should know where these metrics are displayed; typically, it's the web address of a dashboard. These metrics should be sufficient to tell everyone what the health is of the product development system. They might not tell *why* there is a problem, but they should indicate whether there is a problem and therefore if conversations should be had.

Notes

1. itrevolution.com/the-three-ways-principles-underpinning -devops/
2. www.zapata.com/recherche-developpement/

8

Product Design and Agile 2

There are a great many ideas and practices in the community of product design, worthy of many books. This chapter is not about product design. This chapter is about Agile 2 ideas that might help in defining one's approach to product design and how product vision, design, and implementation might best link together, from an Agile 2 perspective.

Agile Ignored Design from the Beginning

In a January 2020 talk,[1] John Schrag, Autodesk's director of experience design, talks about his experience working under Lynn Miller (mentioned later in this chapter) at Alias during the 1990s. He explains that they had an effective design process that was well integrated with development, but then in 2001 the Agile Manifesto was published. Alias brought Jim Highsmith in to Alias to train developers. This revolutionized their software practice, but Agile had no mention of user experience (UX), and the design team suddenly were "at ends," as Schrag put it. He went on to say the following:

> *"We would do a usability study, but by the time we were done, development would be three phases off; or we'd be designing something and they'd have changed direction in one of their Agile sprints: we couldn't give them feedback when they needed it, and we kind of got disconnected. It was horrible. And the funny thing is we were going to other design conferences and seeing the same problem among other people . . . ."*

Early Agile basically instructed developers to collaborate with customers, but the manifesto's authors failed to acknowledge that product design is a substantial activity in its own right and vital to successful outcomes.

Apple is famous for having a design studio that operates independently of the rest of the Apple company.[2] At Apple, product design is everything. Apple does not care if it adheres to Agile or to any particular philosophy except for its own, which has proven to be successful. This illustrates just how important product design is—it is arguably more important than Agile.

Product design is a critical activity that requires focus and its own iterative and experiment-driven process, just as product implementation does. Yet, the two are highly intertwined: to know if a design feature is practical, one must have the help of the technology teams to determine if they can build it for a reasonable cost.

> **Agile 2 Principle:** Product design must be integrated with product implementation.

It is also sometimes the case that there are technology standards that need to change to make a design concept feasible. Steve Jobs famously forced AT&T to change its voice message protocols so that the iPhone Voicemail feature could operate visually, where one can see all of one's pending messages. In that case, design goals not only influenced the product, but changed a major partner's product as well, and it took Apple's CEO to make that happen.

Technology Teams Need to Be Equal Partners

Product ideas often originate from new technological capabilities. The most ubiquitous example of this is probably the laser. After it was first created at Bell Labs in 1960, it was often described as a solution in search of a problem; today, lasers drive the Internet's fiber-optic cables and are used for printing, for 3D additive manufacturing, and for a thousand other uses.

A more recent example also pertains to telecommunication networks. Each time a new protocol generation for cellular data is developed, telecommunication hardware providers around the world launch new products.

Autonomous vehicles are yet another example. The invention of "deep" neural network algorithms and the availability of low power multi-core GPUs has made a whole range of driver-assist features possible.

> **Agile 2 Principle:** Product design must be integrated with product implementation.

So, the arrival of new technologies enables the creation of new product features or even entire new product categories, which flips the common notion of a business visionary who comes up with an idea and then asks technologists if they can build it. Instead, a common scenario is that a new technology arrives, and people ask, "How can we leverage that to create game-changing products?" That process is technology-driven, not business driven, but it still needs to be informed by a "jobs to be done" thought process, because otherwise the product ideas will be off the mark.

This is why product designers—product visionaries we should say—and technologists need to work together. They are two sides of the same coin in a technological society. One does not drive the other; they drive together.

The Product Owner Silo

In the way that Agile methods are practiced in most organizations, each development team usually has a Product Owner (PO). That is a role defined by Scrum, but it is ubiquitous—although not universal—among Agile teams even if they use something different from Scrum.

It is worth pointing out that the Agile Manifesto says nothing about a Product Owner. In fact, one could argue that the Agile Manifesto's authors would frown on such a concept, given that the manifesto repeatedly mentions the customer, rather than a customer proxy role such as a Product Owner. Yet the Product Owner role is extremely common among Agile teams, due to Scrum's popularity.

The concept of the Product Owner is that one individual has decision-making authority about a product's vision and features. That single point of accountability solves a major problem: a team knows who to ask if they are unclear about what a feature should do or the feature's priority.

There are several problems with the PO role, however. One pertains to how it is often implemented, and others pertain to the role itself.

Since the PO is responsible for the product vision, that implies the PO has an intimate awareness of the users—has empathy for them and understands their "jobs to be done." Yet, in large organizations, often a separate marketing group works with customers. The PO is often a business stakeholder or a business analyst who is assigned the task of creating the product. As such, they did not play a significant role in defining the vision and are not intimately connected with actual users.

It does not have to be that way, and it often is not. Many POs have deep knowledge of the business problem that the product was conceived to solve and are perfectly capable of defining the product's vision. That is how it should be.

> **Agile 2 Principle:** Any successful initiative requires both a vision or goal, and a flexible, steerable, outcome-oriented plan.

Another problem has to do with assumptions about the PO's availability. According to the Scrum framework, a PO is supposed to be very present with the team—available for questions and discussion. But a PO who has intimate knowledge of the business is almost always someone who has significant operational responsibility. That means that building the product is only part of their job.

What if there are many teams? For a complex product, there will be several, maybe scores of teams. A single person will not be able to spend significant time with each team, and if they have operational responsibilities, they will have even less time available.

So, the PO role concept is inherently unscalable, defines a role that is a proxy and a single point of failure, and also embeds an unrealistic assumption that the PO can dedicate a large portion of their time to the product development effort.

What usually happens is that a PO assigns a subordinate PO to each team. These subordinate POs are usually business analysts, so they understand the business domain. The question is, were they involved when the product vision was defined? If not, they should have been.

Another problem is that there are often many stakeholders. The product's users are one stakeholder group. There are often others, including those who sponsor and fund the product development, those who will operate it from within the organization, and possibly other affected parties or groups responsible for oversight of products. Each of these has a say in the product's feature set and its implementation. Thus, one PO individual cannot autonomously speak for all of those independent interests.

In that situation, most organizations create a stakeholder team and appoint one person to act as its spokesperson—the official PO. The stakeholder team defines a process for reaching decisions, which the PO is expected to convey and stand behind.

This arrangement can work, but there are things that can go wrong. For one, it is often the case that real users are not participating stakeholders. If the user community is not governed by an individual— say, the users are the public—then there is no individual to represent them on the stakeholder team. Regardless, users need to be included to obtain their actual feedback about the product as it is developed. A methodology for achieving that needs to be applied—participatory design and/or dual-track design and development (discussed later in this chapter)—some approach that embeds frequent user input into the product's effectiveness, from a user's perspective.

Another thing that can go wrong is that development teams act as order takers, instead of having first-hand interactions with real users or being included in the product design process. As a result, the developers do not develop a tacit feeling for how the users will be using the product. That lack of tacit feeling greatly handicaps developers and inhibits their ability to have insights about product functionality and how to fine-tune it.

The Agile Manifesto did not mention design, although it did mention users (implicitly, as "customers"). The manifesto did not have an "order-taking" approach to product development. It recommends "Customer collaboration over contract negotiation." It says, "Our highest priority is to satisfy the customer." It states, "Agile processes harness change for the customer's competitive advantage." It also says, "Business people and developers must work together daily throughout the project." If one interprets "customer" as the user, then these all scream for collaboration with real users.

But somehow, the PO role has supplanted the customer. Too often, Agile POs dictate features as stories, the user (customer) only sees the product at release time, and the development teams have zero interaction with real users.

That is not only an enormous dysfunction, but it is a distortion of the clear intention of the Agile Manifesto.

The Need for Early and Frequent Feedback

There is a big difference between someone saying that they like something and their being willing to actually use it—let alone purchase it.

People tend to only choose things that they need (or *think* they need). There is not enough time in the day to try things that one does not actually need. Unless they are on leisure time, people look for shortcuts. They look for quick solutions. They generally will only invest the time and energy required to obtain and learn a product enough to be able to use it, if they have a burning need—an emotional or utilitarian need—and they think the product might address that need.

That is why liking a product is not enough. They have to *really want* the product.

The implication of this is that surveys about product demand are not very reliable. People answer a survey based on what they think, but it takes little effort to check a box in a survey—far less effort than what is required to obtain and learn to use a product. Just because someone says they like a product does not mean that they will actually use it when and if it becomes available.

Agile 2 Principle: The only proof of value is a business outcome.

To know for sure, you have to engage with them and actually give them the product. They have to actually use it. Then and only then can they assess if it was worth their time.

Also, only by using a product can one experience it and discover if there are things about it that are off-putting or conversely things that are surprisingly delightful.

The dilemma is that to enable someone to try a product, you have to build the product. That sounds like a catch-22, but what you can do is give them access to an inexpensive proof of concept. The proof of concept must have the delightful features, but it can be missing other things. It also need not be made robust or in an economically reproducible way. In other words, it need not be "production ready." It can be assembled using off-the-shelf components. For software, it can be built using scripts, instead of a maintainable programming language, or even more economically using wireframes.

Agile 2 Principle: Obtain feedback from the market and stakeholders continuously.

Agile 2 Principle: Validate ideas through small, contained experiments.

The selection of trial users matters. Are they early adopter personalities? If so, they have a lower bar for being delighted by a new product. However, you might need to test mainly with early adopters first, because they tend to be more willing to overlook the rough edges of a proof of concept. Alternatively, you can pay or incentivize ordinary users to serve as beta trial users.

If using early adopters, it is important to remember that early adopters are not a reliable indicator of how a product will be received by a more diverse population. You therefore need to calibrate the response of the early adopters against the profile of acceptance that you would expect for a distribution of user types.

Trial users will provide extremely valuable feedback about features: what they like, what they do not like, and features that they think would be valuable but that are not present. The product team can then iterate, by adjusting the product and sharing it with a new group of trial users. This iteration should continue until there is confidence that the product will be well received by the market.

Don't Just Provide Features— Solve Problems

The traditional approach to product design is to determine the features that users need and define those as a set of product requirements. The development team then builds a product that meets those requirements.

That approach works fine for something that is not used directly by people. For example, one might use a requirements-based approach for a mechanical engine part or for a microservice. It is not a good approach, however, for anything that is used directly by humans. Humans are not machines, and when they use a tool, they are not simply executing a repeatable action; they are applying a craft.

When a human uses a tool, they are applying the tool to achieve an objective. It is therefore really important to understand the objective. If one has a deep understanding of what users of a tool are trying to accomplish (their "jobs to be done"), one can design a much better tool.

That is why the UX approach to product design centers around understanding the user's experience and what they are trying to accomplish, to inform what kind of tool they need. From a UX perspective, designing a product is not about defining requirements; it is about conceptualizing products that help someone to do a job or to be more effective or to achieve a goal—or to be delighted.

Amazon uses an approach that it calls working backwards.[3] In that approach, a prospective product manager writes a press release for a hypothetical product. If the press release is found to be compelling, the product gets built. During product development, if features get proposed, they are added only if they are implied by the press release; otherwise, they are considered scope creep. In that way, the press release serves as the definition for the product's minimum marketable feature set.

All this does not mean that collecting requirements is not important. It is—even for UX design. Users often expect certain things to come with a product. For example, they usually expect it to be reliable or easy to repair. Those are important requirements that need to be documented; otherwise, product development teams will overlook them.

There might also be social goals that the organization has. These cannot be merely stated as requirements; social goals need to be internalized by everyone in the organization. One of the Agile 2 team worked at Google and says that during his time there, Google's motto and Code of Conduct statement, "Don't be evil," was ever present in people's minds and served as a guide for everything they did. (The phrase has since been removed.)

In a Medium article, "It's Time for Tech Companies to Adopt a 'Do No Harm' Approach," Michael Moskowitz, CEO of Moodrise and former chief global curator at eBay, writes,

> *"We believe that the cultural values of Silicon Valley urgently need to be transformed to truly embrace integrity, decency, transparency, diversity, and social responsibility alongside growth and the bottom line."*

Just solving a user's problem might not be enough: one might desire to solve it ethically. For example, at a major bank, one of us heard an executive describe credit card holders who pay off their balance each month as "deadbeats"—flipping the normal meaning of the term, because from the bank's perspective, those were the "bad" customers— the ones who did not pile up debt that accrued interest. An ethical company might have a different perspective.

To understand users, the best way is to involve them directly. Talk to them. Embed with them if you can. Create a proof of concept and then give them access to it.

For internal business products, such as information systems that enable back office operations, the users are office personnel and other systems. To understand those users and their work, one must not only talk to them, but one must also study their business—their data flows and the values streams that they support. It is usually not necessary to perform a detailed analysis; that is time-consuming and overkill. What is needed is to understand their work and their challenges, and for that purpose, understanding the primary activities and flows of data and value are essential. The details do not matter initially and can be filled in later. What matters is achieving understanding.

It is not usually enough to merely automate people's work, however, because much work has a human element. It is often better to seek to understand how to use automation to make people more effective. We discuss that in the next section.

Products also need to provide feedback to the product team, which reveals how the product is used. Such "telemetry" features need to be built into the product. Data on reliability, feature usage, performance, and other aspects that ultimately affect the users or that impact product lifecycle issues or feature popularity are extremely valuable. Those requirements need to be enumerated during product development, and data analytics features documented along with feature acceptance criteria.

> **Agile 2 Principle:** Data has strategic value.

> **Agile 2 Principle:** An organization's information model is strategic.

One thing we feel strongly about is that organizations should take responsibility for their products. Too many products today are insecure or untrustworthy. A survey of open source software developers revealed that most feel that "securing their code is a soul-withering waste of time."[4]

> **Agile 2 Principle:** Product and service offerers should feel accountable for the impact of product or service defects on their customers.

Thus, the culture of developers does not prioritize security, and it falls to organizations to make security a priority. Too often when there are security problems, customers are left with the problem, and the creator of the product bears no accountability.

Participatory Design

A product design approach arose during the 1980s in Scandinavia and eventually became widely known as *participatory design*.[5] Arguably, the core idea of participatory design is that products should *empower* users, not replace them. It rests on the premise that users need to see products as tools and that what users do is not reducible to mere data flows.

The methodology of participatory design is to engage with users to develop the product design, rather than asking them what they want and then separately creating a design for their approval. This participatory process helps the professional designers to capture what users are actually trying to do and what enablers would help them.

An important element is the emotions involved. For example, do users view certain features as "great" rather than merely "necessary"? The goal is to put the designer in the space of the user and have an equal dialogue and engage in a process of co-creation.

Participatory design is somewhat opposite of the Taylorist view that work can be decomposed into steps, which can then be refined by a "scientific" process expert, for subsequent execution by workers. The participatory design view is that work involves many undocumentable elements—judgments or craft skills—that cannot be reduced to rote (and presumably automatable) steps.

If product designers must immerse themselves in the world of the user, what about product implementers? That is, do the engineers, programmers, or craftspeople who build the product need to also immerse themselves with users?

If one believes that what happens after design is mere execution, then that implies that what product developers do is decomposable into steps. But it is not: there is an immense amount of judgment and iteration that goes into product development and implementation. We have also already explained that

Agile 2 Principle:

Product development is mostly a learning journey— not merely an "implementation."

ideas frequently originate from engineering—not just from the business side—because technology presents opportunities that business might not be aware of.

This strongly suggests that product implementers should have immersive involvement with users. That way, the implementers will "get it" with respect to what the product design is trying to do. They will see a sketch for a product layout, and say, "Yes—that is where we are enabling the user to do *this*, so that *those* can be verified and they don't have to worry about *that*." This fills in a lot of blanks in the developer's mind, and they will also think of alternative implementations that they can suggest to the business and the users, based on the developer's awareness of what the technology can do.

It is not usually practical to immerse all of the developers with the users for long periods of time, but a compromise is often viable. For example, suppose that the design team spends a few days with users once every two weeks. They can take two developers along—each time two different ones. That way, the developers experience firsthand what the users do, how they do it, and what their perspective is. The developers will learn what really matters to the users and what does not matter. This is useful information for developers.

Single-Track and Dual-Track Approaches

Having a design team work together with implementation teams is challenging because the process of design is complex, and it is best done by a small team; if one includes an entire technology team or teams, the process will bog down. That is why the "dual-track" approach was devised by Lynn Miller and her team at Alias.[6]

In that approach, design operates as its own activity. However, it intersects on a regular basis with development. The design team works by itself, perhaps using a process that includes users, but also with the participation of at least one member of the development team. In parallel, the development teams work by themselves, but with the participation of a member of the design team. The design team is one "track," and the development teams are another "track."

The two tracks, or perhaps a few members from each, collaborate on a frequent basis, perhaps cadence-based. The collaboration produces product feature implementation requests, as well as design requests. After each collaboration, the design team resumes its independent work, and the development teams resume their work.

The design team and the development teams each have their own backlog, which must be refined and prioritized in some manner. However, the backlog is not the totality of their work: they each have sufficient free capacity to brainstorm and experiment with their own ideas. That free capacity is important, because it is where inspiration comes from. Those ideas are then exchanged during the next collaboration. Socratic leadership can be used to help make the most of the free capacity.

An alternative version of the dual-track process is more linear: the design team comes up with designs and then on a regular basis feeds those designs to the technical team, which implements them, perhaps as a demonstrator or perhaps as a production-ready feature. That process is more appropriate for more mundane needs, where one is not trying to change a user's behavior or improve how things work but instead one merely needs to implement a known set of needs in the most effective way.

A dual-track design process need not be one of these two binary methods but can blend some of each. Even if a mostly linear process is chosen, the development team can still be given a little time to explore their own ideas. How much is a matter of judgment.

The dual-track approach contrasts with a single-track approach, which is the norm for Agile teams today. Many organizations use a process whereby a set of requirements, perhaps expressed as stories, is fed to a development team, which might include a product designer on the team, either part time or full time. This might be a UX person or a business analyst. The team works collaboratively to devise solutions that meet the intent of each story.

The single-track approach is essentially like the second variant of the dual-track process, except that the design activity is merged with the implementation activity. However, the designers tend to work on individual stories, which affords them only a granular and incremental view, so this approach tends to produce designs that are less holistic.

No approach is better than another. Each has their place. Choosing the right approach is a matter of judgment and should be driven by the nature of the product and the jobs to be done of the users.

Notes

1. www.youtube.com/watch?v=q2dRk3hokEw
2. www.fastcompany.com/3068474/the-real-difference-between-google-and-apple
3. medium.com/inc./amazon-uses-a-secret-process-for-launching-new-ideas-and-it-will-transform-the-way-you-work-aec5c9121ae
4. www.techrepublic.com/article/open-source-developers-say-securing-their-code-is-a-soul-withering-waste-of-time/
5. "The Methodology of Participatory Design," *Journal of the Society for Technical Communication*, volume 52, Number 2, May 2005, p. 164. Available online at repositories.lib.utexas.edu/bitstream/handle/2152/28277/SpinuzziTheMethodologyOfParticipatoryDesign.pdf
6. research.cs.vt.edu/ns/cs5724papers/miller.agile.pdf

9 Moving Fast Requires Real-Time Risk Management

Have you ever noticed how Agile squirrels are? They can leap between branches that are up to 10 feet apart,[1] and they can scurry along the side of a brick building, their tiny claws grasping the edges of the bricks with lightning speed. Their speed enables them to make millisecond decisions, changing their grasp as needed to keep their hold or changing their direction if a tree branch proves unexpectedly unstable—all in time intervals so short that humans cannot track it.

Source: www.pickpik.com/untitled-squirrel-forest-rodent-tree-park-40164

When a squirrel is still, watching, its head twitches so fast that a human cannot see it move—that is how fast they change their focus. Many small animals tend to have fast reaction times,[2] but not all. Worms have slow reaction time. Sloths dwell in trees like squirrels do, yet sloths are notoriously slow-moving. So it depends on whether fast reaction time is important for that species' survival in their particular niche. The same is true for organizations: some need speed to survive, while others do not. Usually speed means *agility*—the ability to rapidly detect that change is needed and respond without delay.

Agility reduces to being able to change one's mind rapidly, in response to something not going as expected—in other words, in response to a newly perceived risk. For a squirrel, if it grabs a branch and the branch suddenly feels like it is not as sturdy as the squirrel thought, the squirrel needs to immediately pivot and reach or spring for a new branch, mitigating the risk of falling. Of course, an implicit assumption is that there is a goal and that agility is not aimless.

Squirrels pay attention to their entire visual field. They have excellent eyesight, and unlike humans, whose eyes focus an image only on a small central area called the *fovea*, the squirrel eye creates a focused image across its entire retina.[3] This is a strong indication that a squirrel can monitor every part of what is within sight—not just one tiny spot like humans.

Agility also requires being able to focus on many things at once. That way, you are never surprised: you always have your attention on *every* aspect of what is going on. The squirrel not only has its eye on the branch that it leaps to, but it has its eye also on every other branch within reach so that if the first branch fails, the squirrel does not need to waste time looking for another—it already has several that it has been watching.

Those who are familiar with the Observe, Orient, Decide, Act method devised by US Air Force Colonel John Boyd,[4] or its cousin Observe, Plan, Do, Check, Adjust[5] written about by many Lean authors, will know that agility rests on continuously and rapidly performing this sequence—always observing and integrating one's observations, always planning, and continually acting and checking.

Agility is about having constant real-time feedback about how things are going, and always having options in mind, so that one is ready to pivot if the real-time feedback starts to reveal problems. Thus, the tactical response capability is extremely fine-tuned.

The Need for Real-Time Feedback Loops

Prior to the Agile movement, the mainstays of management were meetings and reports. Managers have always had calendars full of standing meetings—status meetings for various stakeholders. In addition, reports provided a snapshot of information at a point in time. By the time you received a report and had a chance to look over it, the information was probably at least a few days out-of-date, but in yesterday's slow-paced environment, that was considered timely.

The Agile movement championed short, orchestrated meetings and *information radiators*—prominently displayed metrics and charts that show truthful trends and status—instead of status meetings and reports. That was an improvement in many ways, but there was an unintended consequence: because Agile also emphasizes collaboration across functional groups, managers and teams suddenly found that they needed frequent communication with more parties, so calendars filled up with even more standing meetings.

This unintended consequence was destructive. In his book *Essentialism*, Greg Mckeown relates a discussion with a senior vice president at a major global technology company who told him that he (the vice president),

> *"spends thirty-five hours every week in meetings. He is so consumed with these meetings he cannot find even an hour a month to strategize about his own career, let alone how to take his organization to the next level. Instead of giving himself the space to talk and debate what is really going on and what really needs to happen, he squanders his time sitting through endless presentations and stuffy, cross-functional conversations where nothing is really decided."*[6]

Information radiators were supposed to replace status reports, but it is sometimes difficult to automate an information radiator because the important data might not be available—data such as end-to-end *feature* cycle time, or the rate of *unique* production incidents, or product acceptance by users—and so all but the most standard information radiators generally need to be created by hand. But since communication is now cross-functional, who is making sure that those cross-functional lines of communication are actually happening?

Recall that in Chapter 5 we explained that leadership is needed along the many decision dimensions of a product ecosystem. To make decisions, one needs timely information. In a hierarchy, a manager can direct a particular subordinate to prepare a weekly status report, but today's multi-product ecosystems are not strictly hierarchical, and there is usually no one specifically assigned to maintain the many cross-functional lines of communication that exist for today's complex interwoven and componentized products.

As a result, information radiators that cross hierarchy boundaries tend to not get created unless there are particularly energetic Agile coaches or other individuals who take it upon themselves to do it. Also, if a wall is littered with information radiators, then people will stop looking at them because there are too many, and only those who happen to stroll by that particular team will ever see them, which is problematic if there are many teams.

So while the Agile solution was a good idea, it was not sufficient and did not work at scale.

There is a germ of wisdom, however, in the Agile preference for short, orchestrated meetings and information radiators. The wisdom is to keep communication simple, easy to consume, and visible—or at least easy to find, so that if the culture can be altered so that people look instead of waiting to be told, they will find it.

The problem is that standing meetings are not a general-purpose answer for the need to communicate and should be avoided—even short, orchestrated ones. As we explained in Chapter 6, meetings carry an extremely high cost, so they should be used only by exception, not by rule.

Some standing meetings are worth the cost. For example, a team-wide or program-wide standing meeting might build camaraderie, and that is important. Meetings are also valuable for maintaining key relationships. But if there is a need to regularly communicate about a particular class of issues, too often the knee-jerk reaction is to create a standing meeting about it.

Aside from the few meetings that are needed to maintain camaraderie or a business relationship, meetings should be reserved for two additional situations: (1) when there is a pressing need for multiway conversation, to dive into a complicated issue, or (2) when members of a group *decide* that they want to collectively engage in a creative or collaborative activity in real time. *Meetings should not be used for sharing information*, and decision-making needs to be event-based, not cadence-based.

To achieve agility, one needs to (1) receive real-time feedback on how things are going, and (2) keep all options in view. Having a process for continuously providing feedback and responding (thoughtfully) to it makes the system *self-correcting*. And to be able to respond without delay but thoughtfully when a correction is needed, one must be continuously evaluating options in a holistic manner.

How can we create real-time feedback loops but avoid standing meetings and avoid the problems of information radiators? And how can we keep our list of options always up-to-date and always analyzed?

In other words, how can we create a finely tuned tactical response capability that is always accounting for strategic goals?

An important fact is that management needs to become more tactical. It needs to operate in real time. Management needs to create or get created (through the creativity and initiative of their teams) a tactical response process that is always measuring, always evaluating, and always synthesizing and informed by strategic goals. And the process needs to cross hierarchy boundaries.

To look at the problem this way is to look at the product ecosystem as a system. This is *systems thinking*, which we described in Chapter 3. Systems thinking is about looking at all of the pieces of a system and attempting to develop an understanding of their interactions and how they behave collectively.

The Scrum framework endorses *empiricism*, which is an experimentalist's view. Empiricism is a philosophy that emphasizes using real-world outcomes to inform one's approach. Systems thinking relies on empiricism for validation, but the goal of systems thinking is to obtain *understanding*. If one has understanding, one can better predict what will happen if one makes a change.

The DevOps movement strongly embraces systems thinking, because the core goal of DevOps is to view the entire product development and delivery chain as a single complex process, instead of as many separate processes with handoffs. DevOps also embraces feedback loops, as the primary model for ensuring that one meets one's goals. Feedback loops are also an important element of systems thinking.

Agile was intended to empower individual teams to build and ship products, but what happened is that Agile became a silo of its own, confined to the programming and functional testing aspect of product development. This happened because the Agile community failed to address the many levels of product development. Agile frameworks such as Scrum also embrace feedback loops, but in practice the feedback tends to be defined for a team, as a silo, instead of a product being delivered to actual use.

DevOps attempts to revive the spirit of thinking about the whole product and all the steps needed to deliver it to actual usage and end-to-end scope feedback loops. This was the original intention of Agile: to "satisfy the customer."

Creating Real-Time Feedback Loops

How do we create *effective* real-time feedback loops to make our product design, development, and release process self-correcting?

We already explained why you should get rid of most (not all) standing meetings. Instead, be tactical about meetings. We propose that you should always be asking yourself the following tactical Big Five questions:

Big Five Questions

1. What issues do I know about?
2. Who needs to be aware of those issues?
3. What is the best way to resolve each of those issues?
4. Where might there be issues that I do *not* know about?
5. What things do I need to be checking in on?

Agility is first and foremost about dealing with each situation as unique; context is everything. In an Agile setting, each person's job is to be thinking about the five big questions every day. Less often, they should be reflecting on how well things are going, and whether their feedback loops need adjustment.

A feedback loop generally takes one or a combination of these overlapping forms:

- **Validation**—that is, proactively checking something, such as running a test, or performing a manual inspection, either as a test or as a real-world experiment.
- **Information published** or shared about ongoing activities, such as through a dashboard, message, or other means. This can include information about ongoing validation activities.

Metrics as Feedback

We all have information overload, so the worst thing to do is provide so much information that people do not have time to look through it. Information that contains metrics should focus on key metrics. But what are those?

Every product initiative should have metrics: (a) leading (behavioral), and (b) trailing (outcomes), at each level:

- **Product business**—pertains to business operation.
- **Product operation**—pertains to the operational efficacy of the product.
- **Product platform operation**—pertains to the operational efficacy of the platform. The product platform is the infrastructure or shared system within which the individual products operate. (This is distinct from the *business* platform, which consists of the product platform *and* the products.)

Outcome (trailing) metrics pertaining to the product's business might include things such as the rate of conversion from a prospect to a paying customer, the frequency of use of a product feature, product reliability and usability as perceived by actual customers, and so on.

Leading metrics pertaining to the product business might include things such as how many beta users were given access to the product before it was launched and how much

> **Agile 2 Principle:** The only proof of value is a business outcome.

promotion was done to make people aware of the product.

Outcome metrics pertaining to the product's operation might include its availability (*up time*), its performance, its return user rate, its defect rate (as reported by users or recorded as production incidents), and so on. Leading metrics might include test coverage, for each level of the product architecture (that is, not just unit test code coverage); end-to-end test frequency; and design review coverage (to what degree have the many portions of the product received design-level review).

Some metrics cannot be reduced to numbers. For example, there is not an effective metric for test coverage. We are not talking about "code coverage"; we are talking about *test* coverage—that is, the completeness of the tests with respect to the myriad needs of the product. In lieu of a number, what is useful is a level of confidence of the process used to define the test suite.

For example, if using a Behavior-Driven Development process, has a test expert reviewed and enhanced all (or just some portion) of the product's test specs? If so, then the coverage can be considered to be high. If only a portion has been reviewed and enhanced, then the coverage should be seen as less confident—which might be okay for some components or features, but not for others, depending on how mission-critical each component or feature is.

Metrics are an educated guess at what one can measure that might indicate efficacy, so continually reflect and revisit each metric to assess its usefulness. Metrics should be continuously refined, like anything.

Metrics should also cover the strategies of the organization, the platform, and the product. If agility is an important strategy, then you should have metrics that cover it at each level. For example, feature lead time is a strong indicator of product development agility.

Some metrics—including very useful ones—are subjective. For example, "customer satisfaction" is subjective but can sometimes be assessed by a survey—converting subjectiveness into data.

Each leader should be responsible for a set of metrics. Remember that leaders are not hierarchies—each area of concern should have a leader, whether designated or emergent. Individuals may lead in different areas, and individuals may have other responsibilities as well. The important thing is that no metric should be without someone who is responsible for maintaining it—ideally the metric calculation should be automated. If no one is responsible for a metric, then it will not be maintained, and it will become useless, and no one will pay attention to it. People should also participate in the definition of the metrics that they are accountable for.

People Need to Understand the Metrics— *Really* Understand Them

Leadership needs to create (coach, teach) a culture that understands metrics.

Everyone needs to understand metrics. They need to know the concept of "garbage in, garbage out" and that metrics can mislead you. Metrics never tell the whole story and are not a substitute for inquiry and conversation. Metrics do not replace models or understanding. But metrics are essential, because they help you decide if your approach— your experiment—is working.

A metric is not proof. A metric is an indicator. One should never make a decision solely based on metrics, but metrics are evidence, and as such, metrics should be extremely important factors in any decision. Discussions about issues should almost always reference metrics, even if to challenge the usefulness of a metric.

People also need to have some awareness and general understanding of the major metrics outside of their immediate area of work but that pertain to other aspects of the product or platform.

For example, a product manager needs to understand the development efficacy metrics for the product. A traditional product manager might feel that "development efficacy"—that is, things having to do with the building of the product—are technical and are handled by the development lead, but that is a myopic view in an age in which delivery cadence is an important factor for business agility and time to market as well as time to respond to problems reported by users.

Let's be concrete. Suppose the product delivery dashboard reports that a new product build is integration tested once per week. Should that be a concern to the product manager? It should be, because it indicates that the integration test process is not fully automated, so if there is a problem with the product that needs to be fixed urgently, it is unlikely that a

Agile 2 Principle: Technical agility and business agility are inseparable: one cannot understand one without also understanding the other.

Agile 2 Principle: Business leaders must understand how products and services are built and delivered.

fix can be turned around in less than a week—and that delay will affect users.*

Incredibly, it is often the case that technical development managers do not fully understand their product development processes. The reason is that the processes that are most effective today have only come to be available to most organizations over the past decade, due to the availability of commodity virtualization and on-demand cloud services, so development managers who obtained their experience before that have not experienced the new methods and often do not realize just how different these methods are.

> **Agile 2 Principle:** Technology delivery leadership must understand technology delivery.

> **Agile 2 Principle:** Technology delivery leadership and teams need to understand the business.

One of us had a series of conversations with a development manager at a Fortune 100 company, in which the manager still retained the mental model that code "moves" from one environment to another through phased testing—a 1990s model at best—and that is no surprise because that is when he obtained his formative experiences. And this manager was a key player in the company's most advanced cloud-based systems, due to his position in the hierarchy.

Metrics can play a role in helping to educate managers who have legacy mental models. If that manager had been told that development teams were publishing their process metrics on a dashboard, and the dashboard included end-to-end test frequency and build frequency, that legacy thinking manager might have started to understand the continuous nature of today's development methods.

Even better, provide DevOps and DevOps-aware Agile coaches to talk to those managers and help them to learn today's methods. Once people learn the new patterns, they start to identify with them and become advocates.

Technology leaders—indeed, all tech team members—also need to understand business metrics that pertain to the product and platform. They cannot live in their own technical bubble, paying attention only

---

* This is actually a complex issue, and pertains to the sufficiency of "component contracts" for defining component semantics, and the confidence in component contract tests as sufficient for covering component interactions and the entire system's temporal behavior.

to the up time and response time of their product. They need to understand not only what the product is used for but what its critical success factors are. How else can they have a sense of responsibility for those aspects of the product? How else can they pay attention to those aspects and perhaps think of better ways to use the technology that might advance the product's success?

If a tech team member sees a business metric for the product and does not understand that metric, that means the tech team member does not understand the way that the product's success is being judged—and that is a huge red flag.

Better Information Radiators

The information radiator concept was a great idea, but it does not really work for complex products and large organizations.

One challenge is that leaders often are leery of publishing the true state of their organization in a dashboard that all can see (Agile dashboards are presumed to be viewable by all within the organization). However, transparency is a cornerstone element of being Agile. Without radical honesty, there can be no informed discussion about issues, and if there is not a culture of "letting it all hang out," that is, telling the full story without hesitation, then it will also almost certainly be true that there is a culture of concealment, blame, and posturing. Establishing a culture of honesty, supported by blameless inquiry-based discussion, is one of the most important things that leaders can do.

Information radiators also have a scaling problem. If one has multiple teams, or if teams are remote, then dashboards and other information radiators have to be online—there is no alternative. The problem is that online dashboards don't get looked at by most people. One does not "walk by" an online page— one must proactively visit the page.

Agile 2 Principle: Good leaders are open.

The challenge then is to make dashboards more visible and easier to find. And mention them all the time—*use* them. Leaders should base discussion around the metrics. The metrics are not the goal, because *outcomes* are the goal (conditional on acceptable behavior), but the metrics should represent what current thinking is about how one can move toward one's goal.

Importantly, there should be a few dashboards that everyone knows about: the dashboards that show the "One Metric That Matters" for each team or group as well as for the product and for the organization as a whole. We will explain that more in Chapter 12.

Also, if the organization is consistent about which dashboards exist, then people will start to look for them in "the place that they should be." Each "thing" made or managed by the organization should have a home page, and that home page either should be the "thing's" dashboard or should have a prominent link to it in a standard place, such as the top of a sidebar menu. Then, if someone wants to check on "how that thing is doing," one merely has to search for the "thing's" home page and click the link at the top of the sidebar menu.

There is a cultural change here that leaders need to drive. Many stakeholders are not accustomed to looking at dashboards and expect that there will be a regular meeting to discuss metrics. Getting these stakeholders to look at a dashboard instead of attending a meeting will require proactive reminding from leadership. Perhaps instead of a recurring meeting, a recurring "newsletter," containing the dashboard link, will remind these stakeholders to check the dashboard. One can even attach an audio summary for those who do not want to read written summaries that might be attached to the dashboard. Making it easier for people to consume information is important, especially when trying to change behavior.

If products, platforms, and other important elements of the organization's activity are always discussed in terms of their dashboard, people will start using those dashboards. Management needs to set the example and make this behavior part of the culture. People need to learn that they are *expected* to check dashboards.

To be clear about what the previous "things" are, each product is a "thing," as is the product platform. In fact, each level or aspect of the organization's activities should have a home page, and the home page should be a wiki, such that anyone within that group—those people who are responsible for that aspect of the organization's activities—are able to update and maintain the page, with change history preserved.

We say "aspect" because there are often "things" that are not part of a hierarchy. For example, there might be a community of practice pertaining to product testing methods. Such a community cuts across products—across the product hierarchy. Yet, the community represents a set of concerns, all pertaining to testing of the organization's products. If that community is important enough to be identified and named, then it should have community leadership in some form, and there should be a home page for the community, just as there is for each product's development teams.

Dashboards can also be promoted through newsletters for a group. For example, the previously mentioned community of practice for testing could have a newsletter about testing methods. In that they can mention their home page and dashboard and refer to testing efficacy metrics that are listed in the dashboard.

Dashboards need thought. They need to be designed so that they are effective—just like a product or anything. If they are treated ad hoc, they risk not being used as effectively as they could be. That is why each aspect—each group or area of activity—needs to be *responsible* for maintaining a dashboard with metrics about their activity. It is up to leadership to establish that as an encouraged *and* expected behavior.

People Need to Read, Write, *and* Converse

Dashboards are not the only form that feedback takes. Feedback can also occur through direct communication: speaking, writing, and reading.

We explained in Chapter 6 why reading, writing, and conversing are all important skills when working with others. It used to be that writing was considered a critical skill for business. Somehow, today's culture downplays writing, and it is no longer shocking to hear someone—even an executive—say things like "I don't write. I just call a meeting."

Why then did our college education invest so much of our time in learning to write well? Were all those term papers for nothing? Perhaps the "three Rs" should be just the "two Rs"—reading and 'rithmetic — and we can all forget about writing? Except that if no one writes, then there is nothing to read, so the "two Rs" become the "one R." However, today there is a groundswell claiming that people do not need to learn much math either, so maybe it is the "no Rs." That sounds like a great dumbing down of civilization to us.

Agile was partly a reaction to *too much* writing such as big documents created by requirements analysts, architects, and consulting companies. The Agile movement said, "Forget that—let's just have a conversation."

But that went too far. Big heavyweight documents are not useful. A reliance on documents conflates information (what is on paper) with knowledge (what is in someone's head), losing sight of the fact that documents must be explained, and there needs to be an opportunity to ask questions and have discussions. Knowledge is the result of reading, thinking about, conversing about, and restating information that has been received. Knowledge is a model that one has built in one's head, in response to information, deep thought, and dialogue—written or spoken. The relationship between information and knowledge is well described by the "Data, Information, Knowledge, Wisdom" pyramid.[7]

Writing is still an important part of that, particularly for complex issues.

If one needs to get a quick answer to a question, then a text message or email is the least intrusive way to do that, because the person you are asking is not interrupted. If you are willing to interrupt them, then the best way for *you* to get a quick answer is simply to ask them in person or to make a phone call.

But if you need to discuss something complex, then writing down some ideas might be a good way to enable others to first read your basic thoughts and have time to consider them before having a conversation. People are different in that regard—some will want to read first, and others will want to jump right into a conversation.

> **Agile 2 Principle:** Foster diversity of communication and diversity of working style.

Not everyone can write well, but all of us who have had an education must be expected to write well enough. When writing, be concise and to the point. Be clear and explicit and avoid redundancy. Use an active rather than passive voice. Use few pronouns. And use a separate paragraph for each idea.

Writing is a feedback loop if it provides information or analysis, or both. Others can then consider what you have written and ask themselves extensions of the Big Five questions that we listed earlier.

1. What other issues do I know about, pertaining to this topic, that the sender might not know about, in addition to the issues that the sender has raised?
2. Who else needs to be aware of any of these issues?
3. What is the best way to resolve each of those issues?

4. Where might there be issues that the sender or I do not know about?
5. What things do I need to be checking in on?

If the exchange is occurring via email, you might respond, including some others and being careful to only include those who are directly affected—at least until you determine the full scope of impact. Or, you might "CC" others just as "for your information," providing awareness. That is what "CC" is for; too many people think that if they are "copied" on a message, it means that they must respond. A "CC" is just a heads up—nothing more.

As people exchange their thoughts in email, it will become apparent when people are ready to have a discussion about one or more of the issues. That is the right time to schedule a meeting. Such a meeting has a purpose: to discuss the issue and reach some consensus on what to do about it. It should be a deep discussion that gets to the heart of the matter. Otherwise, those in the meeting will be wasting their time. A meeting should not be about an "ask"—it should be to elucidate an issue and achieve shared understanding. *After* that has been achieved, it should be about reaching agreement on appropriate action.

After the meeting, it is often helpful to quickly document the decision made or the action items committed to and place this documentation in the group's preferred collaboration tool (email, web documents, Teams/ Slack, etc.), so the shared understanding can be referenced, since people's recollection will degrade over time.

Conversation is also an important skill. By discussing the importance of writing, we do not mean to diminish the need for conversation. We simply are pointing out that the power and usefulness of writing has been dismissed in some Agile circles, and that has disempowered those who prefer to share ideas in writing. Despite that, verbal conversation is essential.

Getting to the point quickly and being sparse with one's words are important. When speaking to exchange information or ideas, one should always question whether one has already expressed one's idea and if further words are needed. If no, then just stop and give others a chance. It is also important to try to interpret someone else's meaning, rather than their literal words: people often do not articulate their meaning well. Seeking intention rather than parsing someone's sentence is a huge part of communication.

Writing, reading, and conversing are all critical skills for effectiveness in an organization that deals with complexity. Everyone in such an organization needs to have basic proficiency with all three modes of communication, even if they prefer one over another. Writing concisely is a key skill that needs to be encouraged and championed. Effective conversation also needs to be championed. Socratic discovery, courtesy (letting others finish and not interrupting), and speaking with brevity: these are all important skills. *If it is noticed that someone needs help in one of these modes of communication, they should be coached or offered training.*

If consensus is reached, then the meeting completes the feedback loop. If consensus was not reached, then some reflection on question 3 of the Big Five—"What is the best way to resolve each of those issues?"—needs to be revisited.

Always be thinking about how best to resolve an issue. Do not "knee jerk" to scheduling a meeting or any other default form of communication. Be tactical and smart.

Agile 2 Principle:
Foster deep exchanges.

Agile 2 Principle:
A team often needs more than one leader, each of a different kind..

Validation and Experimentation as Feedback

Our little friend the squirrel validates its choice of a branch when it grasps onto it. If the branch feels as steady as the squirrel thought it was, then its choice has been validated. If the branch turns out to be dry and dead—and snaps!—the squirrel immediately pushes off for another that it has been considering. Our little squirrel has performed a real-world, or production, test of its branch choice.

There are other ways of validating, though. The squirrel could inspect the branch. Does it look dry or dead? If there are tiny offshoot branches with leaves sprouting from the branch, then it is not dead, and it can likely support the squirrel's weight. That would be a design-level validation: the branch choice is verified—with some level of certainty—based on its observable characteristics and a theory of what might cause it to be unstable.

The squirrel could also perform a canary test of the branch: it could put a portion of its weight on the branch but not fully commit to it. If the branch proves solid, it can then fully commit; if not, it can pivot and remove its weight.

Finally, the little squirrel can perform a nonproduction test by looking for a similar-looking branch on the same tree, nearby, and trying that. If the two branches seem similar in every way that matters—according to the squirrel's mental model of what makes a branch stable or unstable—then the test will reveal if the actual target branch is likely to be stable.

Each of the testing approaches is an experiment, a tentative commitment to the actual target branch or to a similar branch that "models" the target branch. The squirrel—if it values its life—will not fully commit to a branch until it has validated the branch, either through testing or through some other means, such as careful examination.

Agile 2 Principle: Validate ideas through small, contained experiments.

Each validation provides feedback to the squirrel about the branch in question. The squirrel then acts on that feedback.

Squirrels are very agile, so their methods for validation of branch choices are extremely adept and well honed. They use the same basic process again and again every time they move from branch to branch. Agility requires a refined and rapid validation process so that a pause is not needed to evaluate each choice—each new product feature. Validation needs to be fluid and automatic and triggered immediately.

Product development is like that. Imagine if each time a new version of a product is created—a tentative branch choice—it must be arduously tested, taking a long time by the development teams. That introduces a long delay before the product version can be released to users. Further, if it turns out that the product version has defects, then it cannot be released, and all that time was wasted.

Now imagine that once a product version is created, the tests start and are performed automatically. Meanwhile, the teams start on the next, improved version. Suppose that the tests find a defect. The defect can then be corrected in the next, improved version, so no time was wasted. Now further imagine that the tests operate quickly and so they can be rerun as soon as the new and improved version is complete.

Since the tests run quickly, there will be little delay, and the product can be released to users after a short period of time.

This parallelism and continuous automated validation (in the form of testing) are key to what is referred to today as *continuous delivery*. In the continuous delivery approach, development never stops; it continues while automated tests are run, and those tests are considered to be *part of* the development process. Also, each time a feature has been completed (but not fully tested), a new product version (incorporating the feature) is passed to the testing process. Versions that pass all tests are considered to be usable by users—to be "potentially releasable." They are the output of the development (and test) process. Thus, the development process produces a continuous series of potentially releasable product versions, like sausages coming out of a sausage machine.

As you can see in Figure 9.1, the development process creates a continuous stream of product versions. Those that have passed all their tests are considered to be potentially releasable to users. Those that did not pass all their tests are not considered to be releasable.

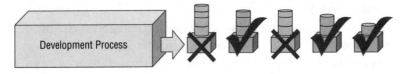

Figure 9.1: Continuous builds being delivered, some potentially usable, some not

Rather than a long development phase that produces a product version containing all of its eventual features, followed by a long separate test phase that validates each feature, the product is developed and tested feature by feature. Each product version is subject to a rapid suite of automated tests. If the product version passes the tests, it is considered to be releasable in a *technical* sense; that is, it has been assessed to not contain defects.

The period of time between successive product versions can be days, hours, or minutes, depending on the number of teams involved, how their automation is set up, and what is being built. For software products, potentially releasable product versions are usually produced many times a day, while for hardware products, the interval might be a few days or weeks, but these are merely typical times—not rules.

The preceding description is highly simplified, and there are some problems with it (which we will get to), but it makes the essential point: that modern development is continuous, rather than separated into phases. Also, the preceding description treats all the activities leading up through the testing of a product as a single "development" activity, which for a real product, that "development" activity would encompass product envisioning, feature definition, and all of the design, construction, and test activities needed to produce a working implementation. Testing might include not only technical testing but actual usage by real users who can provide feedback about how well they like product features, possibly through the rate at which they use those features, which can be collected via real-time usage telemetry.

Putting those complexities aside for this discussion, construction and testing happen concurrently, in different steps of a nonstop "pipeline" of activity, like a car on an assembly line, with each step being done by another task and a stream of cars being produced. This is illustrated in Figure 9.2. The difference is that in an automobile assembly line, each produced car—each product coming out of the last step of the assembly line—is the same as the one before it. But in a development pipeline, each product coming out is slightly enhanced from the one before it, because the inputs to the pipeline—the code or product definition—is continuously being enhanced.

Figure 9.2: Simple pipeline

On paper, a development pipeline looks like a waterfall flow, but it is not like that, because the pipeline is continuous. At any moment, there is a product version being built and a slightly older version being tested, at the same time, and new and improved versions are continuously exiting the pipeline.

To know how well such a pipeline process is running, publish metrics that reveal things about the health of the pipeline. How often are defects found? How frequently are new product versions produced? Are there some steps that seem to be slower than others? Most important of all, is anyone or anything frequently in a situation of *waiting* for something or someone else? Waiting is the big flow killer. It is one of the Lean "wastes." Enabling processes to continue in parallel, without ever waiting, is central to continuous delivery.

Pipeline metrics show how well the pipeline process is running. At the beginning of Chapter 7, we described four levels of issues for product development; pipeline metrics refer to product development workflow and product development technical practices: that is, how well the product development process is running. Pipeline metrics do not show how well the product is performing in any sense. That is the job of product performance metrics, which should be defined both at a business outcome (value) level and at a technical level.

The validation steps of a pipeline, which usually consist of tests, need to be effective. Otherwise, one has tested a product but the tests are not actually indicative of the product's quality. For tests to be effective, they need to be sufficiently complete, and they need to have correct inputs. The primary inputs to tests are test data, meaning tests verify that the product behaves correctly in response to test data.

If development teams must create the test data, how do they know that it is realistic? We mentioned in Chapter 2 that one of us once heard one development lead say to another, "My tests always pass because I create data that I know will pass."

Often business stakeholders are in a better position to define or provide realistic test data, but they do not take responsibility as they should.

The issue of test sufficiency is an important one. In the testing world, this is usually referred to as *test coverage*. It is common in product development organizations that test coverage is managed in an ad hoc way, or not managed at all. Yet managing test coverage is the key to ensuring that one's pipelines are working.

Agile 2 Principle: Carefully gather and analyze data for product validation. (In other words, business stakeholders should provide test data.)

In the realm of software products, test coverage is often substituted with code coverage, but the code coverage metric applies to extremely granular tests and is not very useful for today's complex, highly distributed, multicomponent, real-time microservice-based systems. Test coverage for such systems needs to be managed at all levels—not just the micro level of unit tests.

Flaws in the Pipeline Model

One problem with the pipeline approach is that some steps can be faster than others. When that happens, one ends up with a queue forming in front of some steps. To avoid that, one can "balance" the pipeline, by improving the speed of the slower steps, so that each step has the same speed as the others. One way to do that is to add some parallelism to a slow step. This approach can be used for any steps in the pipeline, whether they are human steps or automated steps.

For example, suppose one step in a pipeline performs a set of tests. Instead of performing all those tests in sequence, additional processing capacity can be added, and the tests allocated among that capacity, improving the overall speed of the test step. This is illustrated in Figure 9.3.

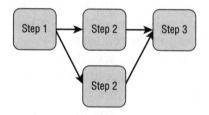

Figure 9.3: Simple pipeline—each step balanced

Another problem pertains to pipelines that share a test environment among a team (or many teams). Such a pipeline is carefully managed and is generally not a good place to diagnose problems, because one cannot "mess with" the pipeline environment. A shared pipeline runs tests, but if a test fails, how does a developer access the internals of the pipeline test environment to see what went wrong? The pipeline is continuously running, and the pipeline script is probably "locked down" to some extent, because it is shared, and so one cannot add steps to the pipeline to see what happens.

Developers can diagnose problems much better if they can run tests in their own workstations. If something goes wrong, they can access the internals, without worrying that they are messing up the pipeline. They can make tentative changes and run the tests again to see what happens. Software developers call that a *red-green cycle*. If developers can create a red-green cycle, they will develop extremely rapidly.

What is really needed is for few problems to be found in a pipeline. In other words, we want the products produced by the pipeline to all be potentially releasable—all having passed their tests. Developers call this "keeping the pipeline green" because it is indicated by all tests showing green in the automated test reports.

To keep a pipeline green, developers must be able to run tests before a new component or product version is sent to the pipeline. That is, the tests must

Agile 2 Principle:
Integrate early and often.

be run locally, on a developer's workstation. Running tests locally in that way is called *shift-left* testing because tests are run upstream—to the "left" of the pipeline. In Chapter 7 we described the concept of "shifting risk left." Shift-left testing is a practice that shifts risk "left."

There are variations of the shift-left testing concept, including running tests in earlier pipeline steps, but the essential idea is to run tests first in an environment or "test rig" where a developer can easily diagnose the cause of problems and use a red-green cycle.

Complex products often have many components, each produced by a pipeline. There is then an integration activity that needs to bring together the pieces produced by the several pipelines.

That integration activity that converges the pipelines is itself a pipeline. This is shown in Figure 9.4. The problem is that the upstream pipelines feeding the integration pipeline are asynchronous with respect to each other. Thus, they are continuously producing components that must enter the integration pipeline. If the integration pipeline is a fixed resource that runs its integration steps on a schedule, then the various pipelines will queue up for it. As we explained, any queue—waiting—is a Lean waste, hindering the flow and increasing overall cycle time through the development system.

A much better approach is to make the integration pipeline resources dynamic, if possible. Thus, each time there is a new component version ready to enter the integration pipeline, the integration pipeline spawns a new set of resources to perform the integration activity, pulling the latest component versions from the other pipelines. This is shown in Figure 9.5.

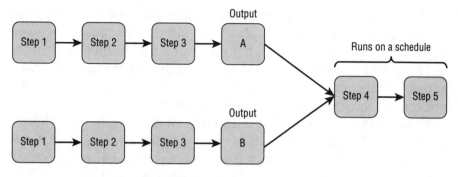

Figure 9.4: Two pipelines—cadence-based

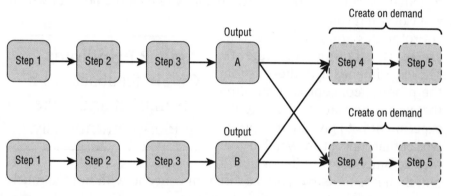

Figure 9.5: Two pipelines—on demand (no waiting)

In that arrangement, the component versions produced by upstream pipelines never wait for the integration pipeline. The integration pipeline produces a continuous stream of integrated product versions, as fast as component versions can be delivered by the upstream pipelines. This maximizes the overall flow of the product development and delivery system.

Feedback: Learn from Product Usage

Just because a product version passes all its development tests does not mean that the product is what people actually want. The only way to know for sure if a product will succeed in the hands of users is to put it into the hands of users.

The problem is, by then it might be too late. You have invested in the development of the product, only to find out that people do not want it, or it is close but not quite what they want, and so they give up on it. This is why it is essential to include users during product definition and give them access to early prototypes so that you can observe and understand the way they perceive the product.

This is not to say that market research is not valuable. In fact, Nielsen research has found that initiatives with strong positive feedback from potential customers before launch are 15 times more likely to succeed in the market than those with mediocre feedback prior to launch.[8] But if you are investing millions or you have only one shot, it is really wise to actually test the market for real before you pass the point at which you can still pivot.

Prototypes and early user trials are market-facing experiments. The experiment's hypothesis is that the product concept is something that users will want to use, and the experiment is to give them a prototype of that concept and validate that they actually do want it—or not. If not, then it is time to pivot to a different concept, with not too much having been invested in the failed concept.

> **Agile 2 Principle:** Obtain feedback from the market and stakeholders continuously.

Experiments that test ideas are an essential form of feedback loop. Experiments should be safe and contained. That is, they should be designed so that if the experiment fails, it does not spell the end of the company. That is, don't bet everything on one experiment so that you have nothing left if it fails. We talked about this in Chapter 5 in the context of black swan events that are unlikely but have a great cost. Here we are talking about events that we can control: the size of our experiment.

Experiment small enough and early enough so that if your experiment fails—and experiments often do—then you still have ample time and resources to try something different. Experiments should be done with care and not be reckless. One should always have a plan for what to do if the experiment fails, just as our squirrel friend has other branches to grab onto if the one it picks fails. Our little squirrel never fully commits to a branch unless it is certain that the branch will hold; it does not want to perform an experiment that could result in "game over."

That said, real-world product usage is unavoidably a kind of experiment. Even if trial users like a product, that is still not a guarantee that the product will succeed; even if the prod-

> **Agile 2 Principle:** Validate ideas through small, contained experiments.

uct succeeds, it might have room for improvement. That is why it is essential to continue to poll customers and try to find out what people think of the product.

The customer is not the only source of information about a product. People who provide customer service or field support are on the front line dealing with real-world users, and they hear all of the complaints. That feedback *must* be collected and provided to the product team; it is a treasure trove of information and is your primary customer-facing feedback loop once the product has reached the broad base of users, and analyzing that feedback is the best way to manage the risk of eventual product failure.

Balance Design and Experimentation

There is an idea in the Agile community that something cannot be fully designed ahead of time: you must try to implement it to discover issues that cannot be anticipated during design.

That idea is not entirely true. We explained in Chapter 6 that it is sometimes possible to fully define a product's requirements up front, but that those cases generally do not include human-facing systems. If one can define requirements up front, then one can presumably perform the product design up front, and it is possible to partially validate a product design ahead of time, by using simulated mocks of a product and having users give feedback.

It *is* also possible to fully create a *technical design* for some things ahead of time, but doing so is expensive. A complete up-front technical design might require developing and using a real-world simulation, which is an expensive and time-consuming process, and the simulation model must itself be validated. The hydrogen bomb was designed and simulated before its first construction and test, and it worked exactly as designed. John von Neumann—one of the most brilliant mathematicians and physicists of modern times—was one of the group that designed the simulation.

Practically speaking, Agilists are right though. For most things, it is just not worth it to try to fully design something ahead of time. It is far more cost effective—and faster—to just build something for real and try it.

But does that mean one does not design at all? Does one just "hack" together something and see what happens, like children mixing chemicals? Or does one do some amount of design? But how much? Will a design "emerge", as we explained in Chapter 4?

Balance between design and experimentation is a key issue for product development: how much up-front design is called for, at each step? The question arises not only at the beginning of an initiative but every time one has to begin work on a new feature. Basically, how thoroughly should we try to design this, and when are we ready to start building?

It is a trade-off: the time and effort required to validate a design prior to building something versus the time and effort lost if the prototype fails (that is, the cost of the failure of a test). The prototype is an experiment; in fact, every construction of an incompletely validated design is an experiment.

The challenge is to optimize this trade-off so that the greatest flow is achieved. If the cost of a failed test is high, then more up-front design is warranted, because it will reduce the likelihood of failure. But if the cost of failure is low, then it makes sense to design a little and then just see what happens. Overall, you want to make the fastest progress, without wasting too much thrown-away product. If your product is software, then there is no "thrown-away product"—there is only the time and effort invested, either for validating or for coding and running the test suite again.

What about when the product fails during actual use? That is, suppose a user is using it and encounters a bug, or suppose the product is running in a server environment, and there is an incident? If the cost of failure *during use* is high, that tends to justify more testing prior to release for actual use, as well as more design effort, according to the trade-offs between design and testing.

What is the root cause of a failure during actual use? And is the cause indicative of a design flaw, an implementation flaw, or a testing gap? As we explained in Chapter 7, any usage or production problem can be seen as a design or implementation flaw or as a testing gap.

The question of where to focus on improvement—on design and implementation or on testing—is a decision that should be informed by the trade-offs between investing more effort in design or more in test development. Would more careful design prevent more design flaws? Perhaps, but it slows development and adds cost.

Would more thorough testing detect more defects? Perhaps, but it slows development and adds cost. This is particularly true if ones uses a proof-of-concept implementation for real: it will likely have all sorts of issues. The question is, are those issues more or less costly than the time-to-market advantage of pushing the proof-of-concept implementation out into actual use, knowing that it was hastily created and not well designed or well implemented?

In practice, one needs a balance between design thoroughness and releasing soon. In Chapter 4 we explained the "hardware-rich" approach used by SpaceX, whereby it performs a robust

---

Agile 2 Principle: From time to time, reflect, and then enact change.

---

design but does not try to perfect it. Instead, they have tools that enable them to rapidly fabricate parts so that they can get new product versions into a test setting as quickly as possible, where they heavily instrument the product and collect data during testing. In the words of NASA administrator Jim Bridenstine,

> "SpaceX brings a very unique capability to the mix that NASA has been lacking . . .SpaceX is really good at flying and testing, even being willing to fail, and then fix, and then fly, test, fail, fix, and they can [iterate] that over and over again very, very fast."[9]

That approach reflects a trade-off in terms of time and effort to perfect a design versus time and effort to learn by trying a change that appears to work. That trade-off is different for each kind of product.

Responding in Real Time

The squirrel's agility arises from its ability to rapidly sense change and then from its ability to rapidly respond. To respond instantly, it must have already tracked and considered alternatives—not waited until it sensed something is wrong with its current path. That's probably why its eyes can focus its entire visual field. If a branch is failing under the squirrel, there is not time to start looking for other branches; the squirrel must already *know* which other branches are closest and that look stable.

The key to a rapid response is to always be thinking about options: What other paths do we have if this one does not work out? What preparations are needed to enable us to take those paths, if we suddenly have to?

Autonomy Enables Faster Response

Teams that can act autonomously can act most quickly. They do not need to wait for a higher-up decision-maker. However, as we have explained, teams are not often autonomous from the beginning. It takes time and effort to make teams autonomous. Within a multilevel organization, authority is necessary at various levels, there are organization-wide or program-wide objectives that need to be met, and there must be assurance that a team will act within those parameters. There also needs to be a support system for when things within a team go awry.

Self organization has implications for team autonomy because in a self-organizing team everyone is individually autonomous, at least at the outset, and yet the team must be able to take direction from above on what needs to get done, and then act in a coordinated manner to achieve that. In a team with an explicit leader, the leader can perform an organizing role, but in a self-organizing team, the team must first decide how to get organized, and then begin working on achieving the goal.

We explained in Chapter 3 that conflict within a team can be very damaging. Teams without training in how to work as a team can easily get stuck in a "storming" phase, in which team members disagree or split into factions. A team needs to be trained and coached in how to operate as a unit and what is expected. This includes how to discuss issues, how to work toward consensus, how to prioritize, and how to treat each other. We have explained that not everyone works the same, so the team also needs to recognize that and establish norms for how they will work collectively while enabling each team member to work in the way that is best for them individually.

One way to address this is to provide formal training, followed by coaching that helps the team to experience working as a group, while also assessing the team's

Agile 2 Principle:

Professional development of individuals is essential.

abilities and identifying potential problem areas (and problem individuals). It might turn out that after a time, the team composition needs to be modified to enable the team to operate as effectively as possible, since some combinations of personalities simply do not work well together.

One can also take a more experimental approach and have a coach work with the team as the team tries to self-organize. The coach can help them to form norms, and interpret issues that arise, while also assessing as described.

Autonomous does not mean leaderless. An autonomous team can have one or more leaders for different aspects of the team's work. There can be an overall team lead as well as leads pertaining to various aspects of work such as technical approach, process setup, and getting (and keeping) things organized. Whether to assign such leaders is a decision that those who set up the team must make in defining the team architecture and the preferred leadership models of the organization.

> **Agile 2 Principle:** Provide leadership who can both empower individuals and teams, and set direction.

Whether a team has explicit leaders or not, autonomy enables fast response, and so achieving a high level of autonomy should be a goal. This requires identifying the kinds of issues that the team might have to deal with and ensuring that they have the training (or coaching) in the skills needed, or support external to the team that they can call on, as well as awareness, understanding, and appreciation of the practices and standards that are important to the organization or program as a whole. It also requires ensuring that the team has learned and demonstrated decision-making ability about those kinds of issues.

Once a team has shown that it can act mostly autonomously, program leadership can articulate goals, and teams will execute in the way that they can best achieve the goal. They then need little supervision and can largely be trusted to deliver.

This does not mean that leadership should just give orders and go away. In fact, leadership should solicit input from teams or team representatives, in the process of discovering and deciding on important courses of action. Leadership should also always be listening to feedback from teams about issues and opportunities. The point we are making is that once a decision is made, an autonomous team can implement the decision without detailed instructions or intervention, which greatly increases the speed of response of the organization.

Agile 2 Principle: Favor mostly autonomous end-to-end delivery streams whose teams have authority to act.

Establish Rapid Coordination and Integration

We have explained how self-organization is something that must be nurtured and that it is not something that necessarily happens well by itself. Leadership is usually needed to prepare people to work in a self-organizing team and to oversee the team since things sometimes go awry or the need changes, which might require altering the team.

The same is true for groups of teams. A group of teams should not be expected to self-organize into a highly functional team of teams. Even if that happens, it can take a long time to occur—possibly too long.

The leaders who establish a group of teams need to develop a strategy that will lead to effective and timely collaboration among the teams. There are various leadership models, and we do not want to prescribe anything in this book; the important thing is that one ends up with a set of teams who act as one. Specifically, they have processes for the following:

- Sharing issues
- Collaborating about those issues
- Reaching decisions, or reaching out to senior leadership when a decision is unclear
- Acting on those decisions
- Measuring how well they are doing, as a group of teams

We will not prescribe an approach, because you might think of better approaches than we can, but we will give some examples.

No matter what approach is used, an important consideration is that *leadership* is needed for the team collective. That is, there needs to be leadership of the group of teams—whether that leadership is achieved through collective action or through an individual who exercises servant leadership. That leadership needs to ensure that the five activities just listed are performed and are effective, and it needs to be accountable. It needs to be able to reach good decisions, in a timely manner, and follow through on those, perhaps using an Observe, Plan, Do, Check, Adjust approach. It needs to identify problems early. It needs to operate in a transparent and open manner, such that well-intentioned approaches that were chosen thoughtfully but turned out to be mistakes are not hidden or punished. It needs to know when to escalate an issue, and it needs to show that it accounts for organization-wide concerns, rather than just the interests of the product development program to which the teams belong.

A Challenge Approach

Talk to the teams as a group—all the members—and challenge them. Tell them that they need to establish processes for collectively performing the five activities listed in the preceding section. Then wait to see what they come up with.

If what they propose is problematic, it might be necessary to engage in Socratic discussion. In that case, you might need to pivot to a leadership team approach, which we discuss in the next section.

Leave them with the power to decide how they will work. Do not object to an approach unless there is no doubt that it will lead to problems; and if their proposed approach must be rejected, explain why.

A Leadership Team Approach

If each team has an individual who is the "lead" for the team or the team's process, then it might be best to bring only those individuals together and form a leadership team for the overall effort. If the teams do not have overall leads or process leads, then it might be necessary to select one from each team.

Asking them to volunteer, or asking teams to nominate, someone risks that those who are most aggressive will manage to arrange to be selected. Often the best potential leaders do not volunteer. As explained by Zhang and others in their *Harvard Business Review* article, "Why Capable People Are Reluctant to Lead," leading can be perceived as risky. According to the article,

> "Based on this research, we found that there are three specific types of perceived risks that deter people from stepping up to lead: Interpersonal Risk . . . Image Risk . . . Risk of Being Blamed . . ."[10]

The article goes on to propose changes to organization culture that can diminish these perceived risks. Yet even if the risks of leading are removed, any kind of nomination or self-selection is often a personality filter, and there is no reason to believe that the filter operates to select the best leaders or the ones who have a leadership style that the organization is trying to encourage.

Instead, it might be preferable to select those who have shown an inclination toward thoughtful discussion, which includes listening, as well as servant leadership. This is why it is important to track who has which leadership style and develop people with the desired forms of leadership in mind. That way, those individuals are known when the time comes to give them a chance to lead. We will discuss this more in Chapter 14.

Challenge them as discussed: to establish processes for collectively performing the five activities previously mentioned. Then see what they come up with, as with the challenge approach.

Agile 2 Principle: Foster collaboration between teams through shared objectives.

A Product Lead Approach

Assign a lead for all the teams—a "lead of leads." Explain to them that the five activities are expected.

This is a traditional hierarchical approach. However, it need not be autocratic. The lead of leads may act as a servant leader, using Socratic and empowering methods and models such as mission command.

Again, we do not recommend using a nomination or voting process to select leads. Being political

An insight that some Agile 2 team members had when discussing multi-team initiatives was:

Insight: Collections of technical teams do not automatically collaborate.

enables one to win others' votes but does not often translate into actual good leadership.

Dependency Management

If there are multiple teams working on a single product, there are inevitably dependencies between the work of those teams. A more extreme situation is when there are multiple products and those products share components. In that case, dependencies exist not only between the work of teams for a product but exist also between the work of teams spanning multiple products.

One often hears that dependencies should be reduced or eliminated. We would say instead that *unnecessary* dependencies should be eliminated, and all dependencies should be identified and a strategy for managing the dependency chosen.

Dependencies are not a bad thing. In fact, dependencies show that one's product (or products) work together to provide more value as a whole. In a technical sense, dependencies are a sign that a system is decomposed into separate components. A key design criteria is that components have high "cohesion"—that is, they each do one thing and one thing well. Components that have high cohesion tend to have logical dependencies on other components that do other things.

There is often confusion about what causes dependencies. Dependencies fall into two main categories: technical and logical. A technical dependency results when one component or product produces a technical construct that is used by another component or product. There are ways to define such technical constructs so that they are somewhat resilient to change. One such approach is to only use "generic" data types in messages going into and out of the component. That helps in some ways but hurts in others. The fact is, any use of something produced by something else is a dependency, no matter how you define it.

There are also logical dependencies. For example, an airline reservation service that lets you change your seat is logically dependent on the service that enabled you to book your flight. There is no way around that. One can try to make that kind of dependency resilient to change, but the dependency is there, because it is a result of the fact that one cannot change one's seat unless one has a flight reservation in the first place. Such a dependency is natural and not a bad thing.

The goal should therefore not be to eliminate dependencies. Instead, in a technical sense, the goal should be to define components that have high cohesion. If one does that, the dependencies that remain will be the right ones—the necessary ones. Once one has components with high cohesion, one can use various technical techniques to try to make their dependencies resilient to change.

Identifying dependencies is a first step, but it only begins there. We have seen groups of teams identify dependencies as a group (of teams) but then go their separate ways without having decided—or even discussed—how they would manage those dependencies.

The simplest way to manage a dependency is to first do the work on the "upstream" component (the component on which other components depend) and then, later, do the work on the "downstream" components (those that depend on the upstream one). That approach works, but it is far from optimal, and it inevitably leads to a phased approach whereby components are integrated in a subsequent "integration phase"—in other words, a waterfall approach.

A much better approach is to define methods for frequently "synchronizing" changes spanning multiple components. There are many technical methods for doing this. They are fairly complex, but integration is inherently complex—there is no avoiding it. The only way to significantly reduce one's product feature cycle time—the time between successive potentially deployable product versions—is to tackle the dependency management process head on.

> **An insight that some Agile 2 team members had when discussing dependency management:**
>
> **Insight:** Explicitly design the dependency management process.

All of the teams need to talk through the chosen approach and refine it over time. For software techniques, one of us wrote an article about the most well-known techniques. The article is called "The Agony of Dependency Management," and it can be found in LinkedIn.[11]

Reduce Lead Time

Product feature lead time is the time from when the people who decide what a product should do ask for a new feature (or change to a feature) and when their request has been met, that is, when that new feature or change has actually been implemented and the new, improved product is available for use.

A short lead time is not a proxy for business value. If one is deploying features that no one wants, no value is realized. Thus, before prioritizing a feature to be built and released, there should have been an effort to validate the idea with users, to the extent that is feasible without actually building it.

The concept of lead time is a general one. One can define it in various specific ways. For example, it can include the time that a feature request sits in a product backlog, or it might only include the time from when the feature has been given priority so that work should start on it, once the development teams can get to it. It might also include the time during which the feature is being considered, and alternatives explored, or the clock might start ticking when the product feature has been precisely defined.

The way that you define the term depends on where you want to shine a light. If some parts of your product conception and development process seem to be lagging, then lead time metrics should include those parts.

Lead time is different from cycle time. Cycle time is a component of lead time but is limited to only the time that a product feature spends in a specific activity such as development, making it more accurately "development cycle time." Since development cycle time is usually what product development people are talking about when they say "cycle time", we will just call it cycle time.

If one is using a "continuous delivery" paradigm, such that development produces potentially shippable product versions, then cycle time ends when the product feature has been completed and validated, that is, it has been added to the product, and that new product version has been fully validated as potentially shippable.

Reducing lead time (and cycle time) increases one's ability to respond to change because one can rapidly respond to changes in the market. If a focus group of users asks for a change to a feature, the time to market of the feature change is constrained only by the lead time. If the lead time is six months, then one cannot quickly produce a modified product and get it into the hands of real users for feedback. Similarly, if a competitor releases a new feature, one cannot add that feature to one's own product any sooner than the lead time.

One way to reduce cycle time, and therefore lead time, is to use automation. For example, we mentioned in Chapter 4 that SpaceX makes heavy use of additive manufacturing—aka 3D printing—to reduce the time (and cost) of producing variations of the metal parts that go into its rocket engines. In fact, 40 percent of its Raptor engine is 3D printed.[12] This enables SpaceX to quickly make design changes, print the parts, assemble a new engine, and test. Making engines is still a lengthy and costly process, but not compared to the labor-intensive craftsmanship methods that are the tradition.

On the other hand, it is important to note that in building a rocket there are many separate pipelines, each with a separate team. There is the engine team, and there are teams for the other parts of the rocket. These teams operate at different paces, and they converge, just as in Figure 9.4. If a problem is discovered in the engine design during a test of the entire assembled rocket, one cannot just print a new engine part and reassemble the whole thing on demand.

Thus, when hardware is involved, a true continuous delivery process is not generally feasible. Instead, a hybrid approach is needed, yet continuous effort to make things more on-demand and dynamic pay off. SpaceX's relentless efforts to reduce its change cycle time has enabled the company to greatly reduce the time to develop new engines and new spacecraft in general, enabling SpaceX to not only be competitive but to set a new standard for the entire industry.[13]

The way that SpaceX builds rockets has changed greatly over the years. It has continued to innovate, always pushing the boundary. To reduce its cycle time, SpaceX implemented an advanced set of software tools that enable the company to design and simulate entire multipart assemblies in real time and then print those parts.

This improvement was made circa 2012—years after SpaceX had launched its first rockets in 2008.[14] Thus, it continued to improve not only its product but *how it makes* its product,[15] focusing on the steps in the process that constrain iteration time.

Anticipate the Skills Needed for Alternative Paths

Product developers will tell you that once they know and understand what it is that needs to be built, it takes them very little time to build it. In other words, the time required to understand the requirements is generally as long as or longer than the time that developers need to design and often build something that meets those requirements.

This is admittedly a general assertion, but the point is that for product

Agile 2 Principle: From time to time, reflect and then enact change.

Agile 2 Principle: Place limits on things that cause drag.

developers to become productive at creating a product feature, they need to first build in their mind an understanding of the feature. Creating that mental model requires time. It usually also requires conversation. Most product developers will listen to or read what is asked for, then echo it back in their own words, and wait for confirmation that they have stated it correctly.

Important aspects of requirements that are often overlooked are the environment and context. For business software, these areas often include existing services and data sources. If the team members do not have knowledge of those services and sources, they will have to learn about them before they can become productive. If the organization is large and has complex data, then the learning curve can be long.

That is why the Agile method known as Feature-Driven Development (FDD), which we mentioned in Chapter 2, begins with the creation of a logical data model of the organization's information, or at least those parts of the information that pertain to the product being built.

Agile 2 Principle: An organization's information model is strategic.

The same applies for other kinds of products and services. For example, if an engineering team consisting of new hires is asked to build an engine, the team will first need to learn about the way that engines are built in the current company. It will be essential for there to be experienced people with that contextual knowledge on the team, and those who are new will require months to learn and become comfortable with the methods, tools, and standards in the company.

Our point is that learning time is a major cost for agility. If teams are to respond quickly, they need to be prepared, in terms of what they know. If they are likely to be asked to work on a particular kind of product, they need to know the data models, techniques, and other domain-specific things that apply to that kind of product in that organization.

This means that time spent getting teams up to speed on things that they will likely work on is time well spent. It also means that time spent creating documentation is time well spent. Documentation of product designs and decisions increases agility because that documentation is invaluable to others who come on board.

We are not talking about heavyweight documentation. In general, a conceptual technical design should include these three things, and no more.

- A set of decisions about the product
- A taxonomy, in which the set of decisions can be expressed
- The rationale for each decision

The decisions are things like "Use third party service ABC for function XYZ." The taxonomy would include "function XYZ." What is it? What is it for? The rationale would be *why* ABC was chosen for function XYZ.

Design decisions can be expressed graphically or textually. For example, a decision to connect two components via a communication link could be documented by a line between those components in a diagram. The point is that the decision is documented, its terms explained, and its reasons stated.

Anything else is superfluous for a conceptual design, and more is not better—having more redundant information or unimportant details reduces clarity, rather than increasing it. Details should be separated into reference documentation or a production design that is linked to or treated as an appendix.

The purpose of conceptual documentation should be to enable others—those who are not familiar with the product's internal structure—to quickly learn how the product works internally and why it works that way and not some other way.

Design documentation is extremely important, unless you have only one team and they will be dedicated to that product for the life of the product.

Agile teams often have a "definition of done" for each component or each product, which states what steps must be completed for a feature to be considered done. To ensure that design documentation is maintained, one might include in the definition of done that documentation should be updated.

Agile 2 Principle: Professional development of individuals is essential—anticipate which skills will be needed by each option.

Agile 2 Principle: Create documentation to share and deepen understanding.

Agile 2 Principle: Favor long-lived teams, and turn their expertise into competitive advantage.

The learning curve that teams incur is an argument (among others) for keeping teams together. Each time a new team is created, it needs to establish a way of working together. Technical teams have a much more difficult time getting started, because they have to set up their tools and automation, which is a large amount of work, and it is unique to each team. They also have to learn the design of the existing system.

We have seen many attempts to standardize team tool setups that are only partly successful, because much of the work involves team members learning about the setup and linking their own ways of working to the shared setup. The tooling setup is also highly dependent on what is being built, so it is not feasible to, say, standardize the setup used for developing a Java microservice with that used to develop a Python microservice, because most of the tools will be different.

Reserve Capacity to Act Tactically

In Chapter 4 we made the argument that some capacity should be reserved to enable teams to perform preventive maintenance and continuous enhancement of their methods, and also to give them some time to reflect on the work. There is another reason to not fully commit development capacity: preserving the ability to act tactically.

Just as governments always find a way to commit every bit of funds that are projected to be available, business stakeholders tend to commit every bit of development capacity that is available through the planning horizon.

The result of this is that flexibility is lost, because if things change, one must take away from some stakeholders to shift development resources to others, and that is always a difficult thing. Few things are more difficult than taking something away from someone once they have been promised that they will receive it. Anyone who has children fears the refrain of "you promised!"

That is why it is important to reserve some development capacity for tactical allocation: that is, for just-in-time decisions about how to use the resources.

In practical terms, this might translate into many things. For example, Scrum Product Owners might be allowed to reserve some future epic-level backlog items as

Agile 2 Principle: Don't fully commit capacity.

"Undetermined" in anticipation of features that might become apparent during development and early user trials. That way, there is room in the plan so that the ability to pivot is preserved.

Notes

1. toddmitchellbooks.com/twelve-things-about-squirrels-that-will-blow-your-mind/
2. www.sciencedirect.com/science/article/pii/S0003347213003060
3. www.wildlifeonline.me.uk/animals/article/squirrel-senses
4. en.wikipedia.org/wiki/Patterns_of_Conflict
5. en.wikipedia.org/wiki/PDCA
6. Greg Mckeown. *Essentialism*. Crown. Kindle Edition. p. 64.
7. en.wikipedia.org/wiki/DIKW_pyramid
8. www.nielsen.com/wp-content/uploads/sites/3/2019/04/setting-the-record-straight-common-causes-of-innovation-failure-1.pdf
9. www.youtube.com/watch?t=265&v=p4ZLysa9Qqg&feature=youtu.be
10. hbr.org/2020/12/why-capable-people-are-reluctant-to-lead
11. www.linkedin.com/pulse/agony-dependency-management-cliff-berg/
12. media.defense.gov/2019/Jun/10/2002143370/-1/-1/0/WF_0069_CASTONGUAY_ADDITIVE_MANUFACTURE_PROPULSION_SYSTEMS_LOW_ORBIT.PDF
13. www.technologyreview.com/2014/03/13/173711/spacex-set-to-launch-the-worlds-first-reusable-booster/
14. www.youtube.com/watch?v=xNqs_S-zEBY
15. www.geoplm.com/knowledge-base-resources/GEOPLM-Siemens-PLM-NX-SpaceX-cs-Z10.pdf

10 A Transformation Is a Journey

Transformation is the new normal. Today, organizations are in a perpetual state of change. We explained in Chapter 3 that the view that an organization exists in a steady state is obsolete and that, today, change is never-ending.

If change is never-ending, that means a transformation is never done! It means that when an organization undertakes a significant transformation and defines a "future state," that if and when the organization arrives at the future state, the future state will be obsolete. This reality means that organizations need to be continually adjusting their goals. It also means that, in a sense, the goal post is perpetually moving away and that one can never reach it. It means that change is ongoing, and *one's organization must be set up to continuously adapt*, always reevaluating, always prepared to pivot its strategy, and always changing.

Small changes that are tactical can be accomplished in a finite period of time. *Some* big, strategic changes can be as well, but it is almost always the case nowadays that when one accomplishes a big change, circumstances have changed, and more change is needed.

Changing an organization's culture is a strategic change. Becoming more agile requires deep cultural change. It is definitely strategic. It is not a "process rollout," because it affects not only *what* every function in the organization does, but *how* it is done. Becoming more agile requires every person in the organization to learn new ways of working. It is transformational, and it is a change that never finishes.

Agile Is Not a Process Change

One of us was involved in a DevOps and Agile transformation initiative in which the person placed in charge of the transformation developed a capability model and used that to plan and manage the transformation. She did this presumably because she had used that approach before for other kinds of change. We explained in Chapter 2 that using a capability model for Agile transformation is not advisable, unless the capabilities are defined in terms of *how* they are each implemented.

One of the line items in the capability table was automated testing. We explained to her that such a capability was meaningless, because it says nothing about how the automated tests must be developed, who develops them, or how they can be assessed to be sufficient. All this fell on deaf ears, perhaps because she did not really understand these issues, not having written automated tests herself.

We also explained that a traditional capability approach was not useful for Agile (or DevOps) transformation, because both Agile and DevOps are not about *what* things you do; they are about *how* you do those things. As in the automated testing example, the question is not whether one has automated tests—all teams do—the question is how complete those tests are, as well as many other parameters that make an automated testing strategy effective. The "how" is entirely what matters. Transformation cannot be achieved unless one addresses the how, and traditional capability models typically list only the what—not the how—although we saw in Chapter 2 that one *can* define capabilities in terms of how rather than what.

It's All About the "How"

It was around that time that the book *Accelerate* was published, in which the authors use a capability model for assessing DevOps transformation, but they define the capabilities in terms of *how* each capability is performed.

For example, one capability is continuous delivery," and in the book it is explained that[1]

> *"In order to implement continuous delivery, we must create the following foundations:*
>
> ■ "Comprehensive configuration management. It should be possible to provision our environments and build, test, and deploy our software in a fully automated fashion purely from information stored in version control. . .

- "Continuous integration (CI). Many software development teams are used to developing features on branches for days or even weeks. Integrating all these branches requires significant time and rework. Following our principle of working in small batches. . .
- "Continuous testing. Testing is not something that we should only start once a feature or a release is "dev complete." Because testing is so essential, we should be doing it all the time as an integral part of the development process. . ."

Thus, the book provides a detailed set of criteria for what makes each capability sufficient to be considered valid. In other words, it specifies minimum criteria for the *how*.

Defining capabilities for transformation is fine—as long as the how is covered. The how is what matters.

Agile Is Much More Than a Process

Executives who have had success with transformations such as Y2K are handicapped with Agile, because they have a mental model for success that does not work for Agile and DevOps.

Agile and DevOps transformation are not a process change.

In process re-engineering, experts define a new set of processes, and then workers learn and execute those new processes. This is a Taylorist view and assumes the following:

- Experts can define a standard process that can be replicated across the organization.
- The process rules can be codified so that anyone can merely execute the steps, following the codified decision rules.

That kind of approach can work for highly repetitive tasks, such as assembling an automobile, although Toyota showed that it was not an optimal approach even for that. For creative work, such a transformation approach definitely does *not* work.

Product development is highly creative. It is not a series of assembly steps. Product development steps include activities such as designing, inspecting, finding the cause of a defect, and continuously refining the overall tooling and processes. These are all creative steps that require a lot of expert-level judgment. Product development is akin to a craft in which one never makes the same thing twice—remember our blacksmith example? Product developers also sometimes have to collaborate with experts, such as machine learning model builders or cloud architects, and reach joint decisions on an approach—there can be no standard "process" for that.

So-called waterfall methods for software development attempted to define discrete steps or phases for building software. It did not work, however, and the Agile movement is a result. Agile was a reaction to—a rejection of—prescriptive stepwise methods for creating software. Core to Agile is the insistence that judgment must be applied at every step. Each "value" of the Agile Manifesto has the form "We value this [approach] more than that," but it never says "Do not do that." Instead, it calls upon the individual to make a contextual judgment about how much of "this" and "that" to do, depending on the circumstances.

There can be no codification of whether to do "this" or "that"—it always depends.

That is why each business function needs to define its own Agile processes, guided by an understanding of Agile ideas. One cannot "follow" Agile ideas; one can only learn them and consider them, because Agile ideas are philosophical. They do not tell you what to do.

Teams that seek to define their own Agile processes benefit from having Agile coaches help them. However, it is important to have open-minded Agile coaches, rather than ones who are dedicated to a particular Agile paradigm or framework. Otherwise, teams will be constrained by which ideas and frameworks the coach is willing to consider.

What Works for One Organization Will Not Work for Another

Imagine that an executive in charge of setting up a new Tesla plant wants to define how cars will be made in the new plant. The executive visits a Ford plant and takes lots of notes. The executive then visits a Caterpillar plant to see how its tractors are made, reasoning that tractors cannot be that different from cars, and thinking that a very different kind of product might help to have a fresh perspective. The executive takes a lot of notes again. The executive then visits a Toyota plant, followed by a Ferrari plant, and again finds that the process at each is quite different from any of the others.

Afterward, the executive is confused, because it seems that all four of these companies build their products very differently. The executive therefore decides to pick one and chooses to use what Ford does as a "template": the Tesla plant can start building cars the way that Ford does and then adjust over time as needed.

We hope that you are thinking, *That is a bad idea!*

Making cars is a complex endeavor, and the manufacturing steps involved differ greatly from one kind of car to another. The volume of production also influences the process, as does the workforce culture.

Yet this is essentially what managers do when they adopt an Agile process used by another organization or implement an Agile framework "out of the box."

The reason that importing someone else's process will probably not work well is twofold.

- The nuances of their products and their environment are different, and nuances matter.
- The process is not what makes Agile work. It is the knowledge and experience in the heads of the people that make it work.

The second reason above is the big one. To be able to make the judgments that Agile requires, people have to know *why* they are doing things the way they are. They have to have decided to do things that way. If you adopt someone else's process, then people do not know why they are doing things the way the process says to do them. You would have to teach them the guidance to convey the judgment to them, but that is not possible. Such judgment is only learned through experience.

If you import someone else's process, it might work eventually, but only after people have tried to use the process and have been able to tailor it to their needs—in effect, creating their own process. Until then, the friction and dysfunction resulting from using a process that they do not understand and did not define will create so much angst and inefficiency that there will be a risk that people will say "Agile does not work" and demand to return to how they used to do things.

Another outcome that is common is that the organization responds to the resulting friction and dysfunction by modifying the process so that it "sounds" like Agile but is actually their original process in disguise. Thus, they fail to actually improve anything or increase their agility.

It Is a Learning Journey

Rather than a process change, Agile and DevOps transformations are a learning journey.

Leadership often assume that the development teams are the ones who need to be the subjects of that learning, but actually *it is the senior leaders who are most critical*, with respect to needing to learn new approaches.

Not only are the process change model and capability model approaches that most senior leaders are accustomed to not effective for Agile transformation, but leaders need to learn how Agile and DevOps work. As W. Edwards Deming once said,

> *"The prevailing—and foolish—attitude is that a good manager can be a good manager anywhere, with no special knowledge of the production process he's managing. A man with a financial background may know nothing about manufacturing shoes or cars, but he's put in charge anyway."*[2]

A surface-level understanding of Agile and DevOps can be dangerous and lead to poor decisions that have a large impact. In Chapter 6 we described a common strategic error in which management misunderstands Agile methods and eliminates all testing experts. Yet in the realm of software, and increasingly in hardware, today's distributed and highly componentized systems need a range of testing methods, and having deep test expertise to guide and support a holistic test approach at all levels of a product is essential.

That is why senior leadership are the first who need to be trained and coached in Agile and DevOps methods. To be able to make good decisions, they need to embark on their own learning journey.

This can be frustrating for Agile and DevOps coaches, because senior leadership often refuse to be instructed or coached about Agile or DevOps. They believe that they don't need to know about it, and that time spent learning is a waste of time, for *them*. Yet it is more important for *them* to

Agile 2 Principle: Any significant transformation is mostly a learning journey—not merely a process change.

Agile 2 Principle: Product development is mostly a learning journey—not merely an "implementation."

understand Agile and DevOps in depth than anyone else, because a poor understanding can lead to disastrous decisions.

We mentioned in Chapter 4 that one of us worked for CIO Mark Schwartz of the US Citizenship and Immigration Service during the time that he undertook a well-publicized Agile and DevOps transformation of that agency. Schwartz demonstrated continuous learning. He would read a new book about DevOps or Agile every week and often would buy copies and pass them out. His biggest complaint about his organization was that people were not taking enough initiative to learn.

Schwartz, who was not a programmer, also wanted to experience first-hand what teams were doing, so he sometimes wrote software programs on the weekend, just to attain a feeling for what the work was like. He did not feel that he was above such tasks, and he believed that having a tangible feel for the work performed by his people helped him to make better strategic decisions.

Is that all there is to it? Just read books and build a few toy products? Hardly. But learning about *up-to-date and competitive methods for how the work might be done* are important elements that too many leaders overlook.

What a Learning Journey Looks Like

Agile transformation is really hard and takes a long time—years. It requires that everyone in the organization change how they work, at a deep level, and it needs to be sustained. People need to let go of "process thinking" and shift to a more collaborative and judgment-based craft-like workflow. They all need to learn the patterns that were described in Chapter 9, which enable people to move quickly while managing risk.

Sustained Urgency

In an organization, people have what they consider to be their main job. They will see transformation as something additional, on top of their main job, because their main job is what they are evaluated on. To make transformation work, it needs to be part of their main job, which means that it needs to be part of how they are evaluated.

In the course of work, crises happen. When a crisis happens, people will align around solving that crisis, and when the crisis has passed, there will be a strong tendency to revert to the old way of doing things and let ongoing initiatives such as an Agile transformation fall by the wayside, hoping it will be forgotten about. They want it to pass, like a fad—like just one more initiative that leadership heaped on them and that did not pan out. They might even unfairly blame the transformation for having impeded their ability to respond to the crisis: "If we had not been in the chaos of this new way of doing things, we could have solved the problem more quickly."

To prevent that, there needs to be a sense of urgency about the transformation, and that sense of urgency needs to be sustained. After a crisis, people need to be reminded that the transformation is still urgent and is still part of their main job.

The only person who can create and sustain a sense of urgency is the most senior leader or leadership team. It is only those at the top who can continue to prioritize change, as crises come and go, each crisis having the power to disrupt the transformation if relentless commitment to the change is not expressed by senior leadership.

> **Agile 2 Principle:** Any successful initiative requires both a vision or goal and a flexible, steerable, outcome-oriented plan.

Skipping Levels

One of the things that Mark Schwartz did was that if he wanted to know something, he did not rely on his staff. Instead, he found out who would have firsthand knowledge and went straight to them, no matter where they were in the organization hierarchy. He did not trust status reports because he realized that managers *green shift* messages (described in Chapter 2) as they pass information up the organization hierarchy to please the people receiving the news. He also asked multiple sources to verify that he was hearing a consistent theme. Reaching out directly to obtain an unfiltered and tangible view is the Gemba practice that we described in Chapter 4.

It needs to work both ways. Executives should have an open-door policy, so that people can reach *up*. That sounds perilous to mid-level managers who want to control messaging to preserve a carefully crafted appearance of harmony and success. Managers who want to control messaging have their own interests in mind—not the organization's. This is toxic political behavior that is common in autocratic organizations, because managers in those organizations have found that executive leadership will punish people if they find out that there is a problem.

Instead, senior leadership needs to cultivate a culture of discussion, where those who express an idea that did not win do not lose standing. Instead, people should

---

Agile 2 Principle: Good leaders are open.

---

gain standing by participating in open discussion (either verbally or through writing) and by being transparent about what is working and what is not. *Achieving that culture is entirely in the hands of the most senior leadership.*

Transtheoretical Model

The Transtheoretical Model (TTM)[3] is the most widely used behavioral change model among behavior therapists. The model posits six stages of change.

1. **Precontemplation** ("not ready")—The individual does not believe that change is necessary.
2. **Contemplation** ("getting ready")—The individual realizes that change is needed and has then wondered, "What does that mean for me?"
3. **Preparation** ("ready")—The individual has visualized how they might change and is considering that they might be able to do it.
4. **Action**—The individual has experimented to see if they can actually change, testing the new approach.
5. **Maintenance**—The individual has been able to continue to change and is starting to accept that they can sustain it and that it is a good thing.
6. **Termination**—The change is sustainable, and the individual now identifies with the change.

During an organization transformation, *each person* must go through these stages. Since Agile methods require people to make judgment calls instead of following procedural steps, people need to be committed to the new way of working; otherwise, their judgments will be to do things the old way.

That means they need to have reached at least stage 2, where they accept that "this new Agile thing" is needed, is the right thing, and will not fall by the wayside after the next crisis. In other words, Agile is not going anywhere. Ideally, they need to see Agile methods as something beneficial. Thus, they need to learn about its benefits and how it can solve problems.

Agile coaches are excellent at giving training sessions about the benefits of Agile. Training sessions should be used to establish the core ideas. It will not convince people, though, and it will not get them through stages 2 through 6. After training, most people who did not already believe in an Agile approach will still be at stage 1.

To get them to stage 6, they will have to go through stages 2 through 5.

The hardest part is the transition from stage 1 to stage 2: getting people to realize that change is needed. Once they believe that things need to change—that the current approach to work is no longer effective—you then have their attention. They then know that leadership knows that the current methods are not working. Many people will have ideas for what to change, but they will only voice those ideas if they know that it is safe to do so, that is, that if their idea does not pan out, they will not lose standing for having voiced an idea that was not accepted.

This is why it is essential to establish a culture of safe discussion. Leadership needs to demonstrate this—to prove—that open discussion does not put anyone's career at risk. To do that, leaders need to "go to the people." They need to join the forums used by every level of staff, voice opinions, and ask others what they think. Most people will not respond, but some courageous ones will, and if someone disagrees with an executive in an open forum, that is gold: it is a chance for that executive to reward the person's courage.

Someone with power should *never, ever* openly disagree with someone who has less power. Instead, ask questions: Why do you think that? What about this? What about that? I see—interesting thought— thanks so much for having the courage and initiative to share your thoughts! That is so valuable!

> **Agile 2 Principle: Change must come from the top.**

Then ask the person who disagrees to talk to you one on one, to talk them through the issue and help them to understand your perspective. Either you will convince them or they will convince you. And then thank them again for their willingness to be honest and take the risk of sharing their true thoughts to someone who has power over them. Show them that you respect them for that. As Jack Welch wrote in *Winning*,

> *"Even though candor is vital to winning, it is hard and time-consuming to instill in any group, no matter what size. . .To get candor, you reward it, praise it, and talk about it. You make public heroes out of people who demonstrate it. Most of all, you yourself demonstrate it in an exuberant and even exaggerated way—even when you're not the boss."*[4]

TTM might appear to be a "big bang" change approach, and it is in a way, but it is the reality of how people behave. The model is based on empirical research on human behavior. The *outcome* need not be a one-time change, however. If the "change" includes accepting that change must be continuous, then that new attitude is part of the change, and it establishes a new paradigm for the individual: that to "be Agile," one must embrace endless change.

Once someone has reached a TTM maintenance level in their thinking, they are well prepared to fully participate in an Agile process of incremental and progressive experiments, which gradually move the organization toward agility.

TTM is therefore about individual attitudes, fears, and a feeling of empowerment. It is not about how agile they are: it is about how secure they are participating in developing a new way of doing things. People need to realize that they need to change, and they need to then overcome their fears about it and discover that they can "be the change." That is a huge behavioral transition for someone.

Identify Preferred Leadership Models

There is nothing more impactful in an organization than the styles of leadership that predominate.

Agile methods do not work under authoritarian leadership styles. Under authoritarian leadership, people are in competition, which prevents them from collaborating effectively. They will make judgments based on "optics" instead of outcomes. They will not voice their honest opinion, which leads to ill-informed decisions above them. They will not provide assistance to another team or team member unless something is "in it for them," which greatly hinders the just-in-time cross-team and individual-to-individual collaboration that is needed by Agile methods.

When beginning an Agile transformation, the most important thing to decide is the set of models of leadership to encourage—models such as those discussed in Chapter 5. How will people with those leadership styles be identified? Does the organization's environment reward or punish those kinds of leadership? Are people who use those leadership styles apparent, or are they quiet unsung heroes who go unrewarded and unrecognized?

People who exhibit the preferred leadership models need to be systematically identified, empowered, and rewarded. They are often individuals who do not see themselves as leaders: the organizers, the helpers, the idea people, the doers. These

> **Agile 2 Principle:** The most impactful success factor is the leadership paradigm that the organization exhibits and incentivizes.

are not the same people—leaders are not all the same. But they all need to be empowered so that they can do more of what they do.

How do you find such people? Do you just let them "emerge"? As we have explained, those who emerge as leaders among a group, so that the group would vote for them, are sometimes not the right ones: those who "emerge" might be the ones who are simply able to garner attention or who are the most persuasive.

On the other hand, those who are appointed to a role of leadership are often the most charming and persuasive, the most supportive of the executive's direction, the ones who take credit for others' ideas and successfully deflect blame for their mistakes.

The challenge is to create a filter that sifts out everyone but those who have the right leadership traits.

How can you identify the organizers? They tend to naturally organize things. They cannot help it: it is in their nature. Perhaps instead of asking, "Who should lead this?" ask, "Who is the organizer?" and then make sure that that person is empowered to be able to organize things. Even better, observe it firsthand, but be careful that what you see is true behavior and not a performance.

Or ask, "Who seems inclined to help others when they are stuck?" and then make sure that person is empowered to help others.

Or ask, "Who is the most proactive 'doer' in the group?" and then make sure that person is empowered to propose actions and facilitate getting things moving.

Or ask, "Who is discerning and tends to have thoughtful ideas?" and then make sure that person is empowered to propose ideas that get considered instead of drowned out by the commotion of day-to-day routine.

Think in terms of empowerment, instead of in terms of "bosses."

Everyone needs leadership training, because everyone is a potential leader. Everyone needs to be able to recognize good leadership and bad leadership. There needs to be a culture of good leadership and a shared set of models of what that looks like.

Someone who is an organizer is a natural for serving as an official team lead, but a team lead does not have to be the organizer. If there is an official lead for a team—that is, a person who speaks for the team in discussions with managers—then that person needs to act as a coach for their team to teach the team how to self-organize and act autonomously. The team lead might need help to do that; they might need a leadership coach. The team might also need an organizer to help them to get—and stay—organized.

The team needs to learn how to have thoughtful Socratic discussions, where ideas are vetted logically and calmly, with people taking turns, and with the discussion progressing toward resolution rather than going in circles. The team needs to learn to test ideas through experiments. The team needs to learn which behaviors are not permitted, specifically that competitiveness is strongly disapproved of. Team members need to have their expectations set that while someone might do something heroic from time to time, the goal is a team goal.

Heroic action deserves acknowledgment. However, repeated heroics is a sign that something is wrong. Why were heroics needed? If someone worked all weekend to get a product out, why was that necessary? It might be a sign that work is not being paced or that there are unreasonable expectations.

On the other hand, sometimes heroics are called for. For example, a doctor who spends 12 hours operating to save a patient is a hero, as are the others in the operating room. Some jobs require heroics from time to time—it is the nature of the work. But if heroics seem to be the norm for getting things done, that is a sign that there is something wrong with the way the work is organized.

The kind of heroism that is toxic is when someone allows themself to become indispensable. Someone who has special knowledge should attempt to share that knowledge to remove themself as a single point of failure. Knowledge should not be a weapon; it should be an enabler, and it should be shared. It is up to leadership to make sure that knowledge is shared and not hoarded by a small number of people, and it is up to leadership to establish the sharing of knowledge as a value that is rewarded as a sign of effective leadership.

A self-organizing team also needs to understand that it is expected to be continuously improving how it works and that its challenges and performance need to be transparent. Transparency needs to be viewed as an organization value, and that value needs to be demonstrated by senior leadership. Admitting the need to pivot is an important way for leadership to demonstrate transparency. Mistakes should always be viewed as something that was tried and did not pan out. That is a normal thing that happens in a fast-paced environment, where one does not have time to exhaustively examine one's course of action, because time to market is critical.

Transparency about performance, about issues that have come up, and about lessons learned are all essential. Teams should be expected to publish dashboards with performance metrics and outstanding issues. Management discussions with teams and team leaders should focus on those issues first and metric trends second. The teams see their own metrics so there is no need to dwell on them. The issues are what the teams are asking for help on. Metrics either support or refute the team's interpretation of their challenges, and should be discussed in that context.

If you expect that in this book we are going to lay out a transformation plan or a transformation organization structure, we are sorry to disappoint. Agile 2 is not a plan or a process to follow. It is a set of ideas. People need to consider those ideas and apply their own experience to craft structures and practices that they feel will work.

Define Structures

The purpose of an organization structure is to decompose an organization into groups that are small enough to be governable and effective.

Large groups of people working on a goal together—that is, a large team—generally do not function well. It is too difficult to obtain consensus, and there are too many directions of communication—a combinatorial number. In fact, in a group of N people, there are N × (N−1) ÷ 2 communication paths, one for each pair of people.

We are not going to provide advice on how to structure an organization. That is beyond the scope of Agile 2. What we will say is that one should consider that *hierarchies have a strong tendency to breed leaders who are more interested in climbing the hierarchy than in helping their own teams or doing what is best for the organization.* A hierarchy tends to reward those who manage "optics," and it also tends to promote those who are strong advocates for themselves, rather than those who are good organizers and help their teammates. Yet, compositional structures such as leadership hierarchies, perhaps supplemented by cross-cutting leadership dimensions, are necessary to scale. Resisting the natural tendencies of hierarchies takes a lot of effort and requires putting the right kinds of leaders in place throughout the hierarchy.

Therefore, in defining any kind of scaling structure, be careful to think about transparency and leadership models, including the kinds of leadership that are needed and where leadership is needed, because leadership is often needed for long-standing issues or areas of concern that cut across any hierarchy.

On the other hand, a flat organization has other problems, particularly that there are not enough decision-makers and people have nowhere to advance to, and we explained in Chapter 3 how Google tried a very flat structure and quickly abandoned it. Other models are possible; flat and hierarchical are not the only choices.

Many traditional managers think in terms of hierarchy and equate hierarchy with structure. They think that if they create certain groups and lines of reporting, then they have solved the organization's problems. Structure, whether hierarchical or not, solves nothing on its own. What is needed is that the organization needs to be decomposed into manageable units, and the *right kinds* of leadership need to be applied for the most important issues.

Establish Understanding and Vision

We explained that transformation is mostly a learning journey. Learning implies achieving understanding. For Agile methods to be used, people must understand those methods, because Agile methods do not tell you what to do; they tell you what to consider. To use Agile methods, people must understand what those methods mean and be able to make judgments about how to apply those methods.

What needs to happen over time is that everyone in the organization must develop an understanding of how and why Agile methods can improve things. In other words, people need to develop an understanding of why traditional methods are slow and inflexible and why Agile methods can improve agility, *if applied with good judgment.*

A particularly powerful way to help people to understand Agile methods is Socratic inquiry: ask them to list some of their current challenges, and challenge them to explain the reason for those challenges.

For example, if the organization would like to create products that its customers are more loyal to, ask why customers are not more loyal now. Perhaps the answer will be something like, "Our competitors have loyalty programs, but we don't." You then might ask why that is, and also why they assume that loyalty programs are the only way to increase customer loyalty. Keep asking until you arrive at what people agree are likely root causes of the current dilemma, and also identify assumptions that might be suspect. Agile coaches will recognize this as the "five whys" process.

An organization's operational agility often pertains to how the work is done. A Socratic inquiry of the flow of work can be done as a *value stream analysis.* In a value stream analysis, people describe how work is done, flowing from beginning to end. The goal is *not* to fully document processes; that would be time-consuming and is unnecessary for the purpose here. *The goal is to establish a baseline of understanding* among everyone in the discussion, pertaining to the sources of bottlenecks or other issues and zero in on problem areas.

Once people feel they understand the root causes of their problems, begin to ask them what to do about it. In the previous customer loyalty example (which probably is *not* done best as a value stream analysis), suppose that the root cause was identified as, "Our business areas do not really know our customers all that well." Some might propose the solution that the organization should perform customer surveys. You could challenge that by asking, "Does that really let you know your customer? Are there better ways that bring you even closer to the customer?"

One of the Agile 2 principles is "Obtain feedback from the market and stakeholders continuously." Another is "The only proof of value is a business outcome." Another is "Product design must be integrated with product implementation." You can remind people of these principles and ask how that might adjust what they propose as a solution.

The suggestions will likely be traditional: mostly process-related or proposed new rules and policies. Perhaps some people will suggest organization changes. That is okay. Some people might also say that the organization's culture needs to change in certain ways.

The foundational objective is to develop a shared understanding, through consensus, about the root cause of the organization's challenges. *Following that*, the objective is to devise some strategies for change.

The way you get them thinking of strategies differently than they are accustomed to is by asking why they think that traditional solutions would work, and talking those through with agility and other business goals in mind. Traditional solutions often would work but would be too slow and therefore not promote agility. The Socratic leader should zero in on that, and remind people of Agile principles that might provide an alternative approach.

If all else fails—if people do not propose a strategy that everyone (including the Socratic leader) agrees will solve the problems, then try a "what if?" question, in which you suggest a particular Agile practice, and ask what the impact of that would be. However, be careful about championing an approach, because then it is yours—not theirs.

A leader might need to champion an approach in some cases, but then the leader must accept that people's commitment to it is based on their faith in the leader personally, rather than on the approach. It is a case of a servant leader saying "follow me." People will follow, but if it leads somewhere bad, it is the leader's fault.

Strategies for improvement should always be viewed as experiments and should always include a way to assess if the strategy is working, such as a metric or other objective criteria.

This process of inquiry needs to happen at each level of the organization. Presumably it has already happened at the highest levels, because otherwise there would be no transformation initiative. The executive leadership needs to define strategies pertaining to how to achieve desired organization culture changes, as well as how to develop key capabilities needed for the business strategy—capabilities that have sufficient agility.

Other levels of the organization need to develop their own individual vision and strategy for how they will perform their functions. It is critical that they *select strategies that increase agility, rather than traditional strategies that merely add more control and slow things down*, which is what traditional approaches have taught them to do. Thus, they need coaches present who are experienced in the application of Agile methods to ask the right questions and present the right challenges when ideas are proposed that will lead to less agility rather than more.

This process of inquiry is not something to do once and be done. It needs to become the new way of operating. People need to view their business function as a *system* that they continually reflect on, examine metrics that reveal efficacy, and strategize for how to improve.

> **Agile 2 Principle:** From time to time, reflect, and then enact change.

Teams and any functional group within an organization are no longer operating according to a predefined process; the process must continuously evolve, improve, and track what is needed as circumstances change. Continuous improvement needs to be part of the routine work that teams do.

In fact, it is not only organizational units that need to perform continuous improvement: every dimension of concern needs to do this as well. That is, each dimension that requires leadership also requires continuous improvement. Thus, cross-cutting aspects of work, which we described in Chapter 5, each need to perform continuous reflection on how well things are going and strategize for improvement.

Identify Target Practices

Once there is some level of shared understanding about root causes within a group and some strategies have been defined, the strategies need to be decomposed into actional things. In traditional process change, these "things" tend to be tasks that define a new process or train people in the process. That approach does not work for Agile change. What tends to work well for catalyzing change is to identify target "practices" and then encourage and measure use of those practices. Practices are new behavioral norms. One does not "roll out" a new practice. Rather, one advocates for it and coaches people in how to use the practice.

For example, suppose that a medical team realizes that sometimes an important step is overlooked. A practice for preventing that is to use a checklist so that each step is checked off as it is performed. Rather than mandating that each procedure must employ a checklist, the practice could be described, and each area of medicine could decide how to apply that practice in their work. It is likely that they will apply it judiciously, instead of across the board, and that is what we want: thoughtful application of ideas rather than blind adherence to rules. This approach is advocated by Dr. Atul Gawande, a professor of surgery at Harvard Medical School, in his book *The Checklist Manifesto: How to Get Things Right.*[5]

A key element is that the practices are defined by those who are affected—those who will have to use those practices.

But how do you make sure that the practices are the right ones? If one asks people who are experienced mostly with traditional methods which practices should be put in place to address current problems, they will likely identify practices that use those traditional methods, which are the ones they are familiar with and that have served them well in the past. The answer might be, just as in the previous example of a team considering how to make customers more loyal, Socratic inquiry that shines a light on today's constraints, such as the need for agility.

For example, suppose an organization has been using waterfall methods for a long time and so most people in the organization are experienced with that style of project management. If they are asked which practices should be added to improve quality, they will likely list practices that add steps for design review, code review, and sign-off, as well as additional documentation to serve as artifacts that document quality.

Now suppose that the same group is tasked with defining practices to improve quality, but with the constraint that the practices must also reduce cycle time. They will likely be hard-pressed to come up with an approach, because all of the quality-related practices from traditional project management involve adding more controls—more oversight, more sequential reviews, more steps, and more documentation—and therefore take even more time. They might try to think of ways to speed up these traditional steps, but they are unlikely to propose anything that departs from the traditional approach.

To unlock their minds from the traditional toolset, they need to be educated on other models—the models described in Chapter 9. Armed with those new ways of thinking and with the help of someone who knows those methods well, the group can start to come up with ideas for how to improve quality while also reducing cycle time such as "Shift our controls from static document checking to real time automated checks," "Automate the integration tests," and "Streamline our dependency management."

There are technical practices that arose early in the Agile community, some of which have morphed into newer practices that are generally considered to be part of DevOps. We described some of the early Agile technical practices in Chapter 4.

The DevOps toolkit of practices is large; these practices can all be provided as ideas from which to select. There needs to be someone participating who knows and *understands* those tools and techniques. That way, they can suggest and explain them, but the final selection of practices should be done by those who will have to implement the practices. That way, they will be personally invested in those practices.

Some of the practices need to be metrics and measure how well things are going. We have explained that generally metrics are of two kinds: leading metrics that assess behavior and trailing metrics that assess outcomes. Both are important.

Some practices should pertain to roles. Should each team have a team lead, a tech lead, and so on? Should there be a cross-team tech lead or cross-team product delivery lead? If so, what do those roles mean? What kind of leadership is expected? How are people selected for those roles, and are the roles permanent?

Some of those choices might have already been made by management, but when defining practices, any change should be open for discussion. Roles have a large impact on efficacy, and people should be able to suggest changes to how roles are defined.

Define the Learning Models

Since an Agile transformation is mostly a learning journey, identifying ways for people to learn is paramount.

Training is essential to provide a shared foundation—a set of models and vocabulary. When we say "training," we are not talking about a two-day seminar. We are talking about real training, in which one learns something thoroughly and is tested in their ability to apply their new knowledge, just as in a university course.

But training is only the first step. Training generally does not convince anyone to change their behavior. It does not usually change how anyone thinks. Most often it only gives them something to think about—something abstract, and perhaps pseudo-experiential in the context of a simulation or mock exercise.

Experiential training can be a powerful learning experience. Nonetheless, nothing can replace the learning that happens over time, as people have real-life experiences on the job that help them understand more deeply the theories and techniques they learned in a classroom or video lecture setting.

Also—and most importantly—people need to experience outcomes that have value to them. It is only when you try something for real and receive the benefit that you start to reinforce your approach and identify with it. It is then that you can reflect, "That worked kind of, but it would work even better if we did *this*. . . ."

That is when people start to take ownership of their process and begin to build true understanding of Agile methods and how to apply them. One can only apply Agile methods by changing them and adjusting them. That is the whole point of Agile: that every action is a decision, and your approach always depends on the situation. There is no script to follow, no step-by-step plan. Agile is about applying Agile models in each situation, choosing from among them, and tailoring them as needed.

To progress rapidly through early experiential learning and then through trying things, people need coaches—individuals who are experienced in the methods and who can engage in deep discussion about what to do in each situation as it arises.

For example, suppose a release manager needs to adjust the product release process to be faster. The release manager is concerned about releasing something that has defects or that has security vulnerabilities. In the traditional approach, each release must be "attested to" by a senior executive, and all the required "artifacts" must have been submitted and attested to by designated authorities. How should the release manager change that process to be faster?

There is no Agile release management process that lays that out. Instead, there are DevOps patterns that address risk management at speed. In Chapter 9 we explained the component pipeline and product integration pipeline models. We also explained how test coverage is essential for managing the risk of a deployment. What if test coverage becomes a control so that risk is managed via test coverage as a control?

To implement that approach as a policy, one would need to get cognizant managers to attest that test coverage is a suitable control, and that if test coverage is considered to be sufficient, per a specified process and set of criteria that allow for the uniqueness of each product, then a product version should be considered to be releasable. The control can be defined based on assessment of the product's inherent risk, such that more risk requires higher test coverage.

The challenge then is to define what is meant by *test coverage*, including defining the kinds of tests, how a test strategy should be defined, etc.

Even that approach is somewhat bureaucratic. What if there are knowledgeable people overseeing the release who assess actual risk in real time? That might be an eventual approach, but when moving from a traditional release process to a more Agile one, the first step usually needs to be one that leverages traditional paradigms to some degree. It should substitute one kind of control (documents) for another (real-time assessment of risk and whether the test coverage is sufficient). Any product version that has sufficient test coverage and passes the tests can be released if the business side wants it to be released—*any product version.*

These are the Agile 2 principles in play:

- **Principle 3.3**: Carefully gather and analyze data for product validation. (In other words, business stakeholders should provide test data.)
- **Principle 8.1**: Place limits on things that cause drag [any source of friction, rigidity, or slowness].
- **Principle 8.2**: Integrate early and often.
- **Principle 8.3**: From time to time, reflect, and then enact change.
- **Principle 10.8**: Validate ideas through small, contained experiments.

Principle 3.3 is what makes it possible to have high coverage and trustworthy tests that are robust enough to trust so that one can release if the tests pass. Principle 8.1 is what encourages us to look for ways to streamline the release process in the first place. Principle 8.2 is how we shorten the time to test and therefore the time to release.

Principle 8.3 tells us to constantly reflect on the process and refine it over time, always trying to reduce the critical path to release while improving the robustness of the testing processes. We also reflect on how we treat one another, the kind of relationships we have or want to have, morale, burnout, etc. It's not just process.

Principle 10.8 advises us to not perform "big bang" product releases that have lots of new features, but instead to release smaller sets of features in each release to see how things go. That applies whether we are releasing in a test mode or in a live mode. Principle 10.8 also tells us to treat changes to our *processes* as small, contained experiments.

Everyone needs to know the new patterns for product development and release. Otherwise, people will not understand the entire flow, and it is important that they do. People need to understand the following:

- The preferred leadership models to consider (Chapter 5)
- Collaboration and decision-making models to consider (Chapters 5 and 6)
- Product design practices to consider (Chapters 7 and 8)
- Continuous delivery patterns and concepts to consider (Chapter 9)
- Continuous improvement practices to consider (next section)

Finding people who have broad experience in the use of Agile and DevOps methods is difficult. There are many people with these things on their résumé, but often people's experience has been in a situation where Agile ideas were used but used poorly, and so those people were not able to experience how Agile or DevOps ideas work well. There is widespread misunderstanding about Agile ideas, and there are also divisions within the Agile community about which models work and which do not.

Most people do not have a good understanding of how Agile and DevOps ideas work at scale. Even fewer realize that what organizations like Google, Amazon, and Netflix do is not similar to the Agile or DevOps models that most have experienced.

Finding people who have broad experience with both Agile and DevOps methods is therefore challenged by these realities:

- There are too few of them.
- Those who must interview prospects often do not have the knowledge themselves, so they are not in a good position to assess the candidates.
- The criteria are not clear-cut. There is not a consensus, but rather, there are different schools of thought—some very entrenched.

We cannot solve that, except to encourage people to consider Agile 2, which tries to harmonize Agile ideas and replace extremism with balance and judgment, and which also advocates that people leave their comfort zone and learn about *all* aspects of the work, rather than just one aspect. In any case, these are some of the kinds of people who are needed:

- Leaders who understand, or are willing to learn about, Agile at scale
- Agile coaches—preferably those who embrace Agile 2's open and balanced (not extreme or dogmatic) approach to Agile
- DevOps engineers who know DevOps continuous delivery practices
- DevOps coaches who know the "big picture" of DevOps—how DevOps works at scale for many teams and many products and who can work with program managers to help improve end-to-end processes
- Product engineers and product designers who know, or are interested in learning, *true* agile and continuous delivery practices and patterns

The second and fourth type of person, Agile coaches and DevOps coaches, should be Agile and DevOps experts—true experts, not necessarily consulting company partners who are more focused on selling services, but experts who have themselves played a major role embedded in *multiple* Agile and DevOps transformations. Those individuals can help executives to talk through concerns and craft a strategy.

It is often effective to pair an Agile expert with a DevOps expert so that they learn from each other and synthesize their views, since they each tend to focus on a different aspect of the problem. Do not make the mistake of treating Agile and DevOps as distinct things.

The strategy needs to address the organizational learning that needs to occur. Toward that end, the strategy might include setting up teams of experts who can help other teams, by providing extended team members and coaches for things like continuous delivery and Agile methods. If your organization is too small to be able to support separate teams of experts who can coach others, an alternative approach is to create a community of practice pertaining to each area of expertise that is needed and make sure that it has leadership (emergent or otherwise) that keeps it active.

The "dojo" model is particularly effective for coaching teams of developers. In that approach, a coach works with an entire team, showing the team how to use various Agile and DevOps practices and then advising them as they attempt to apply those practices in their actual work.

Continuous Improvement

Continuous improvement should be an integrated element of every business function or team. Continuous improvement is the "feedback loop" that ensures that processes get better and better over time, instead of degrading over time. One thing that is certain is that a process will not stay the same as time passes. Things tend to degrade over time unless they receive "care and feeding" that proactively maintains them or improves them, and that applies to processes as well as products.

For continuous improvement to be possible, one must have ways of assessing how well things are going, which implies metrics—be they objective or subjective. The metrics themselves should also be the subject of feedback and reflection. Are the metrics revealing what we need to know? Are they being measured in a robust way? Can their measurement be automated and made more real time?

Thoughtful discussion needs to be part of any feedback loop. Data tells us one thing, but interpreting the data, and deciding how to act on it, is a human activity. The entire range of activities of a business function should be subject to reflection and discussion on a recurring basis, informed by metrics and also informed by experience and judgment.

If the work of a team is technical, then reflection on the technical processes should be a central element for reflection. If the work is collaborative, then the collaborative processes should be reflected on and discussed. If the work includes interactions with stakeholders, then the quality of those interactions should be a focus of reflection and discussion.

Discussion should lead to action: better measurement, ways to test ideas, and trials of refinements to the process. Someone who is a natural organizer or tasked with tracking the continuous improvement process should record notes—insights, concerns, and decisions—and present a summary of progress each time the team reflects.

Continuous improvement extends to the transformation strategy. We have talked about identifying preferred models of leadership and defining organization structures, target practices, and learning models. All of those things should be seen as experiments, and all should be subject to reflection and adjustment over time. Tracking outcomes on organization-scale performance in each aspect is essential. How well are people learning? What practices are getting traction? Where are outcomes best? What is the honest feedback, at all levels? These are important inputs for retrospective reflection on the strategy.

Organizational Inertia Is Immense

An oil "supertanker" requires at least two kilometers to make a complete turn and may require more than seven kilometers to stop.[6]

An organization is like a big ship, and trying to change an organization is like trying to change the direction of the ship. An executive might say, "Make it so," but it is the people in the organization who then must follow through. If they are not in agreement or feel that they can covertly resist and that the executive will abandon the idea due to the resistance, then the executive cannot be sure that the direction will be followed in earnest.

But why would they resist? It is simple: an organization is a collection of fiefdoms. Even the most Agile organization is. Whenever there are groups of people, people have investments in their current situation and endeavors, and they do not want to lose their situation or put their endeavors at risk. Their situations includes the status they have, the control, and the pay; and their endeavors include plans they have underway to advance their situation. Change is dangerous, unless it is change that you are sure will enhance your situation. If the change is something you do not understand, it is natural to be fearful of where it might lead and to resist it.

There are numerous ways that people in an organization can resist a transformation effort. These include the following:

- **Wait it out**: Over time, there will be a crisis, and the crisis will take precedence over this new Agile initiative. Then afterward Agile will have been discredited, and things will go back to "normal."
- **Make change superficial**: A manager might have an analyst "map" current processes to Agile practices, thereby "proving" that they are already Agile. This can look compelling on paper, because remember that Agile is not what you do—it is *how* you do it. So mapping current practices to Agile practices is easy, if one is actually performing those practices in a non-Agile way.
- **Passive resistance**: Claim that you are on board but take only tiny steps and try to always shift the conversation to operational business issues that are "more important."
- **Control information**: Publish status reports that show progress but that do not reveal what is actually happening.
- **Be secretly uncollaborative**: Profess to be working with other business functions, in an Agile manner, but secretly undermine them.
- **Align business support for resistance**: Build support for the status quo in key business areas. Usually people who have been star performers in the old way will try this.
- **Embrace a "hybrid" approach**: Use a hybrid approach that claims to employ Agile practices, but that does not really change things in a substantial way, and claiming success with it.
- **Starve the change agents**: Impede or limit resources available to those who are tasked with championing change.
- **Failing to advocate for change**: By simply not spreading the word within one's group, the group members will sense that their leader does not approve of the new approach, and they will silently close ranks.

This is why leadership at the executive level is essential, and why leadership must practice Gemba to see what is actually happening. It takes strong and persistent leadership to go the distance and keep pushing for change. Otherwise, the organization's "ship" will fail to steer. The transformation will become too costly and fail.

You Do Not Need to Build Anything

If an organization's people know Agile and DevOps patterns and have experience with them, they can start using those patterns immediately, no matter which tools they have. The organization does *not* need to build DevOps capabilities from a tooling perspective.

Some tools are really powerful and even game-changing. Cloud services, for example, changed the way that testing can be done. Instead of using static test environments, you can now provision test environments on demand to enable teams to run tests in parallel without interfering with each other, which enables the parallelism that was referred to in Chapter 9.

But you can still use Agile and DevOps methods if you do not have cloud services available, by using fallback techniques. In fact, those fallback techniques are important even in a modern organization, because there are always legacy systems that cannot use the new tools. In an established organization, one almost always has a hybrid configuration.

It is also inadvisable for most organizations to build their own framework of any kind. The cost and effort to build a framework and maintain it is large. Companies like Netflix and Google do it, but they have huge revenue that they can divert to maintain any homegrown frameworks. In effect, those frameworks become finished products that are maintained by dedicated framework teams. Netflix has a huge tools team for maintaining its in-house tools.

Most organizations cannot afford to dedicate people to maintain in-house tools, sufficiently to make the tools viable. Today, if a developer is using a tool and has a problem, the developer searches for their problem on the Internet. Usually, they find the answer pretty quickly. But if a tool is homegrown, an Internet search will not reveal anything, because the only people using the tool are in the company—or worse, left the company! At that point, the developer is stuck and has to call the people who built the tool; but they are likely busy, and so the developer has no quick path to resolve the problem.

The basic lesson is to only build a homegrown tool or framework if you have the resources to turn it into a polished product that is highly reliable in all ways that people might use it and continue to maintain it as a highly polished and robust product. Otherwise, it will quickly become a "house of broken windows." No one will want to use it, and it will be treated as a problem and a legacy tool that should never have been created.

Notes

1. Nicole Forsgren, Jez Humble, Gene Kim. *Accelerate*. IT Revolution Press. Kindle Edition, p. 85.
2. www.brainyquote.com/quotes/w_edwards_deming_672630
3. en.wikipedia.org/wiki/Transtheoretical_model
4. Jack Welch, Suzy Welch. *Winning*. HarperCollins e-books. Kindle Edition, p. 31.
5. Atul Gawande. *The Checklist Manifesto: How to Get Things Right*. Macmillan Picador.
6. www.usps.org/ventura/art-03-3-tankertips.html

11 DevOps and Agile 2

Just as different Agile experts will describe Agile differently, different DevOps experts will describe DevOps differently—often *very* differently. This creates a dilemma in which executives want to issue a directive such as "use DevOps." Such a directive is entirely unclear.

Worse, such a directive says nothing about what the actual outcome should be. Just as *agility* should be a goal rather than Agile, DevOps itself

should not be a goal. But what is the value proposition of DevOps? That is, what things does it provide? Perhaps those things can be goals.

Also, why do we care? Is DevOps part of Agile? Does it relate to Agile 2?

Yes, and yes, and we will explain why. And DevOps applies to product development in general—not just software development.

Before we start, be prepared that we move fluidly between technical topics and softer topics like leadership. This is because there is no "line" between these things. They are all related! It really is

> **Agile 2 Principle:** Technical agility and business agility are inseparable: one cannot understand one without also understanding the other.

not possible to completely separate the technical and nontechnical aspects of product development today.

What Is DevOps? And Why Does It Matter?

These are some of the descriptions of DevOps:

- DevOps is when you integrate product development with product operations, which is why it's called DevOps (**Dev**elopment + **Op**eration**s**).
- DevOps is defined by Gene Kim's "Three Ways."[1]
- DevOps is the same as continuous delivery.
- DevOps is a set of techniques for rapidly building, testing, and deploying new product versions.

All of these are true. What is usually left out is that DevOps is fundamentally a set of techniques that enable rapid development at scale *while also managing risk in real time.* In Chapter 9 we saw how important it is to build risk management *into* one's automated processes to be able to develop rapidly and still be confident. DevOps practices, which are by and large continuous delivery (CD) practices, are techniques for rapidly integrating, testing, and deploying new product versions while ensuring that those product versions do not have serious defects. If one cannot have confidence in what one has built, then one cannot deploy it. DevOps is really about making risk management a real-time built-in activity, usually in the form of thorough automated tests.

Also, because risk management becomes automated and real time, one can achieve great scale, deploying large and complex products or systems of products on a frequent basis. If that is not agile, then what is?

DevOps also includes many practices pertaining to the operation of large and complex real-time systems: monitoring, logging, and rapidly testing and deploying fixes. There are also architectural patterns that are strongly associated with DevOps, including the "12-factor" criteria and the microservice pattern. The microservice pattern is specific to software, but the 12-factor criteria has relevance for hardware products as well, especially hardware that is produced using digital methods.

DevOps authors often cite these *direct* value propositions for DevOps[2]:

- Frequent deployment (Amazon performs > 1,000 deployments/hour)
- Ability to experiment
- Shorter product feature lead time
- Shorter time to recover from technical problems

However, as we have explained, there is more to it than this. DevOps methods, when used properly, provide the ability to manage dependencies spanning hundreds, thousands, or more components through automated tests so that one can deploy frequently with confidence, giving an organization scale and the ability to maintain really complex products.

The ability to experiment is rooted in the technical agility that DevOps methods provide: one cannot launch lots of experiments if one cannot easily deploy and redeploy different versions of something. Shorter lead time is another outcome of technical agility. The ability to rapidly recover from technical problems is also an outcome of technical agility.

DevOps enables one to manage risk and quality with a kind of "dial." The dial is test coverage: increase the coverage, and your risk goes down, although the cost of developing and maintaining those tests goes up. It is a trade-off. This is the real-time risk management that we have described.

DevOps practices include heavily instrumenting products so that one can tell in real time how things are going. That leads to better availability and responsiveness. That is another element of real-time risk management.

The full set of value propositions of DevOps therefore is as follows:

- **Scale**—deploy big stuff and complex stuff that requires many teams.
- **Technical agility**—short time to develop and deploy a new feature, while managing risk: this in turn provides *business* agility.
- **Automated product quality management**—quality (which translates into reduced risk) becomes a "dial" that one can turn.
- **Enhanced availability**—products are "instrumented" and provide real-time monitoring so that product failures are detected immediately and so that product support teams can rapidly diagnose a problem and create and deploy a fix.

These four values provide direct business value. Scaling means that an organization can build more complex products that interact. Amazon's retail platform contains myriad features, all integrated. Amazon also deploys updates to its online features many times a day; such frequent updates for such a complex ecosystem would not be possible without DevOps techniques.

Since Amazon features usually work, Amazon is successful in managing technical risk, even though they make frequent changes and their platform is complex. The Amazon sites are also highly available, which indicates that there are mechanisms for monitoring every product's behavior in real time and for rapidly diagnosing and fixing problems. In fact, since Amazon makes its in-house tools available to the public through its Amazon Web Services, one can see all of the tools that Amazon uses itself for achieving these things.

Notice that lower cost is not one of the value propositions of DevOps. In fact, DevOps methods are likely to *increase* IT direct costs. To understand this, consider the *irrigation paradox*. Water conservation efforts assumed that by increasing the efficient use of water through reducing evaporation and runoff, water use would decrease so that more water would be available for all. What has been found, however, is that increasing water use efficiency tends to *increase* water use.[3]

The reason that this happens is apparently that water saved through efficiency improvements is used elsewhere by those farmers to grow more crops. Thus, rather than reducing their costs, the farmers use more water and grow more, because the water efficiency improvements effectively reduce the cost of water on a per-crop yield basis, incentivizing farmers to make more use of it.

A similar thing happens with the use of DevOps methods. DevOps methods enable an organization to develop and deploy new product features more rapidly. What happens is that organizations tend to do just that: they push more features to their markets. There are no *direct* cost savings. There might be a reduced cost per feature, but the gross cost of product development does not decrease. In fact, it tends to increase. There might be *indirect* cost savings eventually realized through increased product reliability, and there are certainly top-line *values* provided by DevOps methods, including shorter time to market, but direct costs will usually go up.

Tangibly, this happens because all of the systems and tools required to achieve the preceding four value propositions are expensive. It takes a lot of work to set up and maintain DevOps systems. Also, since DevOps methods enable more frequent testing, teams tend to—guess what?—test more! There are savings in that one no longer has to maintain static test environments, or pay for an army of manual testers, but the tendency is for teams to be always testing: always creating environments, running tests, and then destroying those test environments. That is where the speed comes from. No one has to wait for anyone else to be done with a test environment, or to perform manual tests, but it means that at any one time, there might be many different test runs in progress.

Notice also that DevOps is largely about the process of building and supporting a product. Product design is not usually considered to be part of DevOps, even though a healthy DevOps process would contain robust product design; it is just that the product design activities are usually viewed as a separate "layer" or aspect of the process, apart from the DevOps aspect. DevOps also usually does not include customer feedback unless the feedback is through automated means such as feature usage telemetry. These are gray areas, as there is no definitive specification of what is "DevOps" and what is not.

Isn't DevOps Just for Software?

It is true that DevOps methods were developed in the context of software. However, most DevOps practices apply equally to how modern hardware products are made.

Today, when a hardware product is designed, it is often defined using design tools for solid modeling or electronic design. Those design files are software artifacts—just like code. Teams iterate, and there are pipelines.

For example, consider how electronic systems are typically designed and tested today. An engineer makes design changes either to embedded software or to circuitry, and simulates the design in a computer, running test programs against the simulated design. If the tests pass, the design might then be "flashed" onto a read-only memory (for software) embedded within a chip, or flashed onto a field-programmable gate array (for circuitry) using an integrated circuit prototyping rig sitting on the engineer's desk, to create, in effect, a hardware prototype. Test programs are then run again, but against the hardware prototype. That is, to use our terminology, local component testing. Engineers call it "hardware in the loop" testing.

If the component level tests pass, next level integration tests can be run in which the hardware prototype chip is connected to other chips (possibly prototypes as well) on a "board" or subsystem. If the board level tests pass, tests can be run at a system level, comprised of all of the boards of the product or assembly.

Some companies have labs in which the configuration and setup of tests beyond the component (chip) level can be done remotely and programmatically, greatly reducing the integration test cycle time and sharing of test resources.

Finally, if all the integration tests pass, the latest set of design files that passed all tests can be used to create actual hardware for field testing on a recurring basis.

In addition, the product is highly instrumented and sends back real-time messages about its performance using telemetry, an approach known as *telematics*. These messages are collected through a message gateway, fed to analytical engines that monitor all of the products that are in use, and maintain statistics and generate predictions about the performance of the products.

That's DevOps, and it is hardware.

Gene Kim's Three Ways

In a previous section we mentioned Gene Kim's Three Ways. In Chapter 7 we mentioned Kim's "first way," which pertains to systems thinking. His second and third "way" are as follows, paraphrased[4]:

2. Use feedback loops to create a self-correcting product delivery system and to better understand and respond to all customers; and continuously try to shorten and improve the feedback loops.
3. Use experimentation to try things and learn from those trials and also to perfect how one executes through repetition.

These three ways, systems thinking, feedback loops, and experimentation, are, according to Kim, at the heart of DevOps. These principles are clearly not confined to product development, but can be applied to almost any human endeavor. Viewed this way, DevOps is a philosophical paradigm—a way of thinking. It is about thinking of the big picture, about reflection and correction, and about the importance of trial and error and refinement over time—sage advice.

Agile has always recognized the importance of feedback loops, but in a narrower context. The Agile Manifesto talks about involving the customer, and customer feedback was therefore called for from the beginning. Also, many Agile frameworks specify a "retrospective" activity in which a team reflects on how it is working and makes adjustments.

Agile 2 Principle: From time to time, reflect, and then enact change.

The term *DevOps* arose from the realization that product development and product support and operation cannot be performed well as distinct and independent activities. Both development and Operations need to be part of the product quality feedback loop. To provide the enhanced availability promised by our fourth value proposition mentioned earlier in this chapter, *enhanced availability*, those who respond to problems need to have a deep understanding of the product. Otherwise, they will not be able to diagnose and fix problems that occur. It is also essential that the product be well-instrumented so that problems are detected right away, and that impacts how the product is designed and built. Thus, we see that a sharp separation between product development and operations is not possible if we are to deliver on the *enhanced availability* value proposition. The term *DevOps* was created to refer to more integrated development *plus* operations models, whatever those might be.

Common DevOps Techniques

The philosophical view of DevOps is informative and serves as important guidance. However, DevOps also encompasses a collection of techniques—*practices*—and these practices precede the philosophical viewpoint, which was essentially derived after the fact to explain DevOps. The techniques are those of continuous delivery. You can also argue that there are additional techniques, including some product architecture design patterns, that make DevOps practical. Let's look at some of the practices.

You Build It, You Support It

The idea of "you build it, you support it," also sometimes articulated as "you built it, you own it," is that teams that create a product should also support and operate that product. This rejects the traditional approach of having a development team "hand off" a product to a different group to operate, support, and maintain a product. The rationale is that those who build something know it much better. They also have an incentive to design and build the product to be easy to operate and maintain, since they will be the ones doing the operating and maintaining.

That is the theory, but we have seen many instances in which the incentive was not enough, under the pressure of product owners who demanded that the focus of work be on adding more features instead of investing the time and effort to make the product maintainable. This is usually an outcome of the fact that business stakeholders do not understand how the product is built or maintained, and so they are not able to make intelligent trade-offs between adding more features and making the product more robust.

> **Agile 2 Principle:** Business leaders must understand how products and services are built and delivered.

In the DevOps paradigm, development teams are also accountable for any problems with the product. For example, if there are lots of defects reported by users, this reflects poorly on the teams. And if there is a major incident in which customer data is lost or compromised or the product fails catastrophically in some way, it is often assumed that the development teams are at fault. However, it is often the case that business stakeholders were warned that more effort needed to be invested in shoring up the product, but they decided that was not a high priority.

We mentioned in Chapter 8 that there is a major problem today in that data on the Internet is generally insecure. People do not trust any of the software products that they use today. It is generally assumed that anything is hackable, and indeed almost anything is. That is a tragic outcome for technology that has such potential.

This is very much both a DevOps and an Agile issue. As with the Agile movement, security is an afterthought in the DevOps community. The term "Sec", for security, has been inserted into "DevOps", as if that solves the problem.

The reality is that too often DevOps security relies on two things:

- The built-in security of an underlying cloud provider;
- The use of automated scanning tools.

These are not sufficient. Today's cloud platforms are extremely secure, but it is extremely easy to use them in an insecure manner. Scanning tools only find a fraction of vulnerabilities.

DevOps reflects the needs and demands of organizations. If organizations prioritize security, then more DevOps (and Agile) practices will evolve to meet that demand. Organizations need to prioritize security. They need to insist that developers have training and expertise in secure product development, and they need to create incentives so that security is rewarded, instead of only penalized when something goes wrong.

> **Agile 2 Principle:** Product and service offerers should feel accountable for the impact of product or service defects on their customers.

Continuous Delivery of Product Increments

Continuous delivery is what DevOps is mostly about.

Chapter 9 explained some continuous flow models. We saw that with those models, one can rapidly deliver a sequence of product versions, each incrementally improved. We also saw how one can embed real risk management into the flow—not a "sign-off" type of risk management that does not really manage anything, but actual risk management, where we can adjust our level of quality and therefore risk.

Thus, speed is not at the expense of quality. However, quality costs more: add more validation to the flow, and quality will go up, but there is a cost in maintaining those validation checks. In a flow process, quality and cost are trade-offs, and you can adjust them to achieve the right balance.

A Flow of Experiments

The flow approach is a cornerstone of DevOps. Component pipelines create a flow of component versions, and these converge through product integration pipelines to produce a continuous stream of incremental component and product versions. If one treats each product increment as a feature, each describable by an Agile story, then the flow is a sequence of product versions that map precisely to each Agile story as it is completed and identified as done.

Note that *done* means completely done: there is a new *product* version—not just a new component version—that can be distributed for use. That is, any change to any component has been validated to work *within the product*.

In the contemporary Agile way of looking at things, each story is an experiment. We believe that this approach originates from Steve Blank's work on customer discovery,[5] and it has been popularized in Eric Ries's book *The Lean Startup.*

Agile 2 Principle: Work iteratively in small batches.

Agile 2 Principle: Validate ideas through small, contained experiments.

Under that paradigm, a story is a product feature that, once implemented, can be tried and validated through use by real users. The story is a customer-facing experiment, backed by a hypothesis that users will like the feature. Users can use the feature, or not, and they might also be able to provide feedback. Their response to the feature either validates or invalidates the hypothesis and can result in a new experimental story to modify the feature. No product feature should be viewed as "final" or permanent. Everything about a product is always an experiment, subject to change based on new information or actual use. All designs are tentative; all plans are fluid.

If anything can change, then it is best to try to validate things early as best as one can. That implies releasing small feature increments, rather than big sets of features, because early feedback might prevent one from building the wrong thing, which is costly. Thus, each feature is a small, relatively contained experiment.

The concept of a story as an experiment is not strictly a DevOps method; but it is one of the sources of DevOps value cited by come authors, because DevOps methods make frequent customer-facing experiments possible.

Pipelines

We explained pipelines in Chapter 9. Superficial descriptions of DevOps often focus on the pipeline concept, but omit the importance of continuous product level (multi-component) integration as well as other practices.

Most important, the pipeline is not where innovation happens, it is not where product development happens, and it should not be where most test-and-fix iteration happens. Rather, each candidate product or component in a pipeline should be a fairly well-tested, *improved* version.

Test-and-fix iteration should ideally happen *before* a product or component candidate enters a pipeline. As we explained in Chapter 9, the "shift left" approach advocates that development and testing of a feature should happen *prior* to any pipelines, in the product developer's local development and test setup—their "test bench"—because that is where they can iterate fastest. The pipelines are just a fully thorough check.

The challenge for DevOps-oriented teams is that many features are being worked on in parallel, and there are often many components (usually in separate code repositories) affected by each feature, and the changes spanning all need to be synchronized. There are many approaches for that, including feature toggles and feature branch builds, but we will not get into those here.

On-Demand Provisioning

In Chapter 9 we explained on-demand provisioning. It means that each time one needs any resources for any purpose, one "provisions" those at that moment. For example, if one needs to run tests on a new version of a product, one runs an automated script to create the test setup needed, as well as the new product instance to be tested using the latest template for the new product version. The result is a product—the "system under test" within a test environment or test setup, in which one can run automated tests. When the tests have completed, one has a choice: inspect the product to discover the cause of test failures or simply collect the results and destroy the product instance and the test environment.

This is important because one cannot integrate early and often (an Agile 2 principle) unless one is able to perform that integration *without disturbing others on the teams*. If one uses a shared integration environment, then integration must be planned, and many people must stop what they are doing and participate in the integration activity. However, if a team member can perform an integration on their own, whenever they have a need to, and validate the integration by running integration tests, then others can continue working. Thus, anyone can perform and validate their integration at any time. To be able to do that, one needs to be able to create the required integration and test resources on demand.

> **Agile 2 Principle:** Integrate early and often.

To use on-demand provisioning, one needs an automated script or program for creating a test environment and for creating a test instance of the product and placing it within that environment. Scripts or programs for creating a product environment are referred to as *infrastructure as code*, because the environment, which is considered to be infrastructure for the product, is created by code, meaning a script or program.

When we say *environment*, we are talking about the setting in which the product instance is tested. For server software, an environment will likely be a cloud-based cluster or something similar. For an electronic hardware product, an environment might be a test bench containing a set of programmable read-only memories or field-programmable gate arrays or other setup for designing and rapidly prototyping and testing electronic systems. For a machinery product, it *might* not be possible to completely create the environment using automated means. Some of the product's components might be created automatically using additive manufacturing methods. The product might even be assembled by automated means, but placed manually for testing in a test rig. On the other hand, that might be automated as well. The more that is automated, the lower the cycle time will be.

On-Demand Regression Testing

Automating product integration testing, using on-demand provisioning of resources, is one of the most difficult things for an organization to do if it did not set things up this way from the beginning. The reason is that it requires that a realistic product operating environment, containing all of the things needed for the product's operation, can be creatable automatically. The scale and complexity of that can be challenging. Each component also needs to have an automated and robust script or program for deploying the component into that operating environment and connecting the component into the product. (One can also use the real operating environment if one isolates a portion of it for testing.)

Yet it is on-demand regression testing, operating in an integration pipeline manner, that enables an organization to truly reduce feature delivery cycle time in a substantial way. For a product that has many components, most of the problems arise in the integration of those components: simply verifying that each component satisfies a "contract" is not sufficient: component contracts are a very weak way to prove integration. Integration problems are what keep a feature from being marked as done, sometimes for weeks. Also, if the integration testing process is not automated, then it cannot be rerun quickly after fixes are made. Having high coverage automated integration tests is essential.

If the integration deployment and test process is automated and uses on-demand provisioning of resources, then it can also be performed by multiple teams in parallel. They can each run integration tests of their own changes, before incorporating changes from other teams. That kind of staged integration approach greatly increases the efficiency of integration testing.

There is a technique called *trunk-based development*, in which changes to a component are frequently merged to a single version of that component, called the *trunk*. But there is still the issue of integration with other components. The ability of a team to check if a new version of a component integrates well with the latest trunk versions of other components, prior to moving the component's changes to its trunk, is a really powerful way for teams to verify integration as they work, instead of making changes and then letting the integration pipeline find the problems. Google uses that approach. At Google, a *commit* to a component's trunk triggers comprehensive integration tests with the latest trunk versions of other components. Teams can decide if they want the commit to complete before or after the integration tests complete, knowing that if the tests do not pass, they have "broken the product."[6] [7]

The idea of integrating frequently applies outside of software. Say a team is working on a single document, but each person is focused on a separate section. One approach is to use a single shared document and make changes in real time. However, some of the team members might prefer to share sets of changes to their section only when ready. It would be inadvisable for them to go too long before sharing, because they might get substantially out of sync with what others are writing. For that reason, a good practice would be to merge each of the sections on a frequent basis. It is also a good idea to have someone regularly review the totality of the merges to ensure that they flow together well. In other words, there needs to be someone thinking end to end about the entire product, and how well it all fits together, from a product design perspective as well as a technical perspective.

Shift-Left Integration Testing

We explained the DevOps practice of shift-left testing in Chapter 9 and have mentioned it in prior sections in this chapter, so we will not belabor it here, except to say that it is an essential feature of *effective* DevOps, and make a few additional points.

A pipeline is inherently a batch job. It is not interactive. As such, it is inefficient if people are waiting for it. What a pipeline has going for it is that it is a standardized process that can be run again and again; in other words, it is repeatable. However, a pipeline is not the place where one should be finding one's *mistakes*. A pipeline is better seen as a double-check that reruns tests that should have been run before the job was submitted to the pipeline, and it runs additional tests (perhaps a full regression), which were too difficult to run outside of the curated environment types and testing resource levels that the pipeline has access to. If integration testing is performed prior to the pipeline, then the pipeline should "stay green"—the tasks run in the pipeline should always succeed—most of the time.

When shift-left integration testing, one runs integration testing locally, outside of a pipeline. For shift-left integration testing, the requirement for a realistic product operating environment can be relaxed, and a simpler setup substituted. The goal is not to find *all* integration problems but *most* of them, particularly those that pertain to intercomponent interactions rather than interactions between components and their environment. If most intercomponent interaction problems can be found before submitting to the integration pipeline, then the integration pipeline will usually stay green, and work will be much more efficient. That is why shift-left integration testing is a core practice of effective DevOps and is the primary practice for reducing product feature cycle time.

Getting Started

We have said that one should not adopt an Agile (or other) framework as stated but that one should treat any framework or standard approach as a set of ideas to consider and then craft one's own approach for working.

Yet having templates or starting points is valuable if one treats them as a *starting point for deciding* how to work. One kind of basic plan template that is helpful is a standard set of activities for beginning work on a new initiative. This is useful because there are so many steps that often get omitted.

Agile 2 Principle: Organizations need an "inception framework" tailored to their needs.

One must create the teams needed to deliver a product, but the issue of creating teams is but one to be dealt with. There is also the matter of establishing an end-to-end flow. This is not just an Agile issue; it is a design process flow issue, and it is also a DevOps issue, because DevOps-oriented teams have so much more up-front work to do to get started on feature development, and it is not work that can feasibly be done in the course of feature work. It needs focus and intense effort to get all the things in place needed to be able to run tests both locally and in pipelines and to push new product increments from end to end. The entire development and delivery process is immensely more complex than it was when systems were monolithic and merely needed a unit test integration server.

Let's begin with team creation. In their book *Liftoff*, Diana Larsen and Ainsley Nies describe team creation in terms of creating a charter that encompasses *purpose*, *alignment*, and *context*.[8] The purpose includes the product vision (what the future looks like), its mission (the changes to be brought about to realize the vision), and criteria and checkpoints or waterlines for assessing if the vision and strategy are correct.

Alignment includes identifying the stakeholders and reaching agreement on every stakeholder's role and mode of participation. One might also consider decisions about funding to be made as part of stakeholder identification, but Larsen and Nies include that as part of "context." The context is the landscape of organizations, systems, and funding sources or streams (for a product or capability) or identifying a new source of funding (for a project).

One should also identify critical milestones and the high-level architecture approach, which is important for assessing feasibility and the approximate spend rate or time needed to reach a milestone.

These things might be decided in the course of a Lean portfolio process or some other cadence-based process, but ideally, the process is frequent and not annual so that it can make course corrections. Annual funding is generally too slow today and is usually better for making large-grained decisions about strategic capabilities and funding rates for those, rather than for specific initiatives. As Humble et al. write in *Lean Enterprise*, "The use of the traditional annual fiscal cycle to determine resource allocation encourages a culture that thwarts our ability to experiment and innovate. It perpetuates spending on wasteful activities and ideas that are unlikely to deliver value."[9]

Once these things have been established, getting started entails defining the work for the teams to perform. A common approach in the Agile community is to have business stakeholders define a *backlog* of work—a prioritized list of things to accomplish. However, we learned in Chapter 8 that such a process excludes users and product designers. If the product needs innovative design, a better approach might be to create a design team and establish a working arrangement between the design and development teams. Actual beta users might also be included in the process in some way.

If there is existing organization data or APIs that must be used, then the teams need to discover and learn about them. The effort involved in that tends to be underestimated. Complex products often have hundreds of microservices, and those APIs represent a composite information model. The organization should have that model documented, as a domain design. It might be valuable to have a data architect or solution architect explain the model to the teams, as well as the various APIs and message types that might be used. As we explained in Chapters 2 and 9, the Feature-Driven Development methodology begins with a team-wide effort to create or learn about a logical object model of the data. That learning should not be too academic, however, as the best way to learn something is by using it, once you have a grasp of the overall model.

The product teams must develop a shared understanding of the intended high-level product design and technical strategy, to the extent that those exist as a result of early product envisioning, and everyone must begin the process of refining those, or pivoting on them if necessary. Ideally some of the team members played a role in the creation of the initial product design approach and technical strategy. Everyone must learn them, and they need to be documented in a place and form that permits refinement, such as a product-level wiki (such as Confluence), in which *everyone* can post comments, edit pages (with change history preserved), and subscribe to page updates.

The team members also need to be able to add to the design, including adding new pages and diagrams. The design needs to be a living multilevel document and a focal point for discussion about design issues. That is what Agile wikis were created for.

The product teams must also design their process, that is, how they will select tasks to work on, how they will merge changes, which repositories they will use, and how they will document any APIs they create or change. This is where DevOps comes in, because for DevOps-oriented teams, the process is an order of magnitude more complex than it was for early Agile teams.

As part of defining how they work, they need to define a strategy for testing. That needs to cover testing at every level of the product: the unit level, the component level, the product level, the platform level, and any other important architectural level or dividing line within the system. As part of that, they need to define how they will determine if testing at a given level is sufficient. This is often called *test coverage*. Is one test enough? A hundred? Must a testing expert review the tests to ascertain if they are sufficient? How will they know that the tests are enough to cover the risks sufficiently?

The issue of coverage is arguably the most important of all. It is the risk management "knob": higher coverage means more quality and less risk, but higher coverage costs more. The level of coverage is the key trade-off that an organization has for managing risk in a DevOps setting, as we explained in Chapter 9.

The teams must also decide to what extent they will utilize a shift-left testing approach. Shift-left testing is difficult to achieve. To be able to do it, product deployment configurations and test scripts need to be written in a way to make it possible to deploy locally and run tests locally. We are talking mainly about component tests *and* integration tests. For today's componentized and distributed products, integration is where the problems are.

The teams must also decide what their component-level and product-level merge and integration approaches will be. They need to decide what kinds of design reviews or code reviews will be included in the feature level Definition of Done. They need to decide how they will make sure that what they produce is secure. This is a lot to consider, and if one approaches in an ad-hoc way, one will "paint oneself into a corner", and be unable to substantially improve things later.

This is why it is important to have a template for setting all this up at the beginning of an initiative, or for onboarding a new team. The template should not be a plan to execute: it should be *input* to planning about how to set up the work.

People Know How to Do the Work—Or Do They?

The biggest risk that exists in product development is that someone *does not know what they don't know.*

One of the maxims of the Agile community is "trust the team." There is a core assumption in that maxim that a team of product developers know how to do their job, and one just needs to let them do it and not interfere too much.

That seems like a reasonable assumption. After all, product developers are professionals and are hired based on their knowledge and experience. Just as one expects that a professional plumber knows how to do their job, one should expect a product developer to know their job as well.

It's not the same, though. The plumbing profession changes slowly. There might be new materials from time to time, such as when PVC pipe began to replace metal pipe for many applications, but by and large, change in the plumbing profession is pretty slow compared to, say, information technology.

Product developers today are usually dealing with information technology, because today's products are digitally designed and often digitally fabricated; and of course, software products are entirely digital and consist themselves of information technology, for which the pace of change is extremely rapid.

One problem is that people often do not keep up with change. DevOps was an obscure niche in 2010, but in 2020 it is the dominant paradigm for building cloud-based solutions, and therefore every product developer should know DevOps methods. Yet, our experience has been that DevOps knowledge is in short supply.

The rapid pace of change means that one cannot assume that a technical development team knows what it needs to know. Things are changing too quickly. People are often very behind the change curve.

Teams are also often put in a situation in which they are told to build with something that they know little about. For example, there might be data sources, service APIs, or message types that they need to use or connect to.

The majority of Agile teams—those that use Scrum or other frameworks—are typically asked to start work immediately, and it is just assumed that either they know about the tools, techniques, and systems they will be connecting to, or they will find out about these in the course of coding. In our opinion, that is a great way to end up with a set of Rube Goldberg–like systems.

12-Factor App

The 12-factor app pattern[10] is a set of practices designed to enable highly scalable, resilient online services. It includes practices pertaining to code management, dependency management, service architecture, deployment, and service operation.

This is an extremely important pattern set, and these patterns are widely accepted within the DevOps community.

The patterns are conceptual. Even nontechnical people should be able to understand many of them. Understanding these patterns will provide insight into key DevOps ideas and approaches, and so all stakeholders who are involved with development, including Agile coaches, should become familiar with these patterns.

Lean Metrics

Lean metrics are important for DevOps because DevOps views product delivery as a flow, and Lean metrics are flow-oriented.

In Chapter 7 we explained leading metrics and trailing metrics and mentioned a few. Here we will be a little more concrete, in a DevOps context.

Trailing Lean metrics that are often used or tailored for a DevOps context include feature lead time and cycle time. Other quality-oriented metrics that have a flow nature to them include mean time to recover, defect rate, and test coverage. Note that some of these terms are general in nature, but in the following list, we tailor them to apply to common DevOps circumstances:

- **Lead time**—the time delay from when it is decided that a product feature shall be added to the product, to when that feature becomes available to users of the product
- **Cycle time**—the time delay from when product teams begin implementing a feature to when they have completed the feature, with sufficient quality that the feature could be released to product users
- **Mean time to recover**—the time delay from when a product fails during use, revealing a defect, to when the product is restored to a usable state (with or without the defect)
- **Mean time to repair**—the time delay from when a product fails during use, revealing a defect, to when the product defect has been corrected and the product has been restored to a usable state without the defect (We mentioned this in Chapter 5.)
- **Defect rate**—the frequency with which unique product defects are discovered during use
- **Test coverage**—the level of completeness with which a test suite tests the product's features and all the ways they might be used

or misused, to ensure that there are no defects. Test coverage is not necessarily a number, but instead might be a subjective assessment

- **Quality**—metrics or lists of defects, vulnerabilities, or other problems discovered in the product as a result of automated checks, including security checks on the source code or on a deployed instance of the product

DevOps teams also usually use leading metrics that indicate behaviors that are thought to be beneficial. These often include the following:

- **Test frequency (particularly integration tests)**—the number of test runs that are performed per day (or other period)
- **Merge frequency**—the number of merges per day (or other period) into a component's "main" or "trunk" version, indicating that changes made by multiple people are being synced on a frequent basis

Retrospective Covering *All* Topics

In the Agile community, a *retrospective* is a meeting of a team (or teams) to discuss how well things are going and to identify areas for improvement. DevOps teams usually also use Agile methods, and so retrospectives are common for DevOps teams.

It is essential that development teams discuss both their *technical* and *non-technical* processes during their retrospectives, and their delivery metrics should be central to the technical discussion. For example, if they have been trying to reduce their lead time, they should try to identify the bottlenecks in the process. If the product is experiencing a high defect rate during use, the team(s) should examine their design and test processes and coverage to see how those defects are "getting through." If teams that respond to defects or incidents have a hard time reproducing problems or testing fixes,

> **Agile 2 Principle:** From time to time, reflect, and then enact change.

the bottlenecks in those processes should be examined.

None of this should exclude reflection on other non-DevOps issues. For example, issues pertaining to user experience and user feedback should also be considered. The entire end-to-end process of conceiving and developing the product should be in scope for the retrospective, and sufficient people present from each area to be able to determine the root causes and define strategies for addressing those.

A retrospective is not normally used to define detailed plans for addressing problems. The goal is usually to identify the problems, their causes, or steps that can be taken to reveal the causes, as well as approaches for addressing the causes. Detailed plans for addressing the causes are usually worked out with those who have been identified to work those issues.

Data

Data is the Kevin McCallister of both Agile and DevOps. In the movie *Home Alone*, Kevin is the child who is unintentionally left behind when his family goes on vacation. Yet data is arguably the *most important thing of all* in many domains!

Agile 2 Principle: Data has strategic value.

Agile 2 Principle: Carefully gather and analyze data for product validation.

We have already talked about data at length in Chapters 8 and 9 from the perspective of testing, but we want to explain another way that a focus on data has been lost, at great cost, and that needs to be remedied. We include this in this DevOps chapter because the current situation is partly an unintended (and unnecessary) consequence of DevOps approaches that have evolved.

By the end of the 1990s it had become standard practice that organizations would require that an information model be maintained and that each development team work with a data architect to leverage existing models and update them as needed. This created a data modeling bottleneck for Agile teams, and so organizations have by and large bypassed the data model step. In doing so, they have discarded an important step instead of making the step Agile.

Then something else happened. Organizations such as Google realized that at very large scale, SQL databases are a performance bottleneck.

To achieve greater scale, they had to find an alternative, so they created "no-SQL" databases. These remove the performance bottleneck for two reasons: (1) their tables are generally either simple key-value pairs (the "value" might be a complex object), and (2) they offer the option of relaxing transactional consistency.

A common feature of these new databases is that many of them do not require a schema. That is, you don't have to define your data's structure anywhere. The result was that many teams stopped documenting their data.

The failure to document data has resulted in an inability to leverage data as an asset, as we described already in Chapter 2.

What could be done instead is that an information modeler could be assigned to work with each team, as an extended team member. The model could be maintained in real time, as changes are needed. The data modeler then serves as an expert on call to discuss the data and its meaning, and changes to the model that might be needed. The data modeler (or information architect) is a subject-matter expert on the organization's data, just as product teams usually have access to a system architect who is expert on the organization's systems.

An alternative approach is to make the system architect or product technical lead responsible for understanding the organization's data and documenting it (or ensuring that documentation is maintained). That approach is better aligned with today's preference for encapsulating data within each system instead of sharing data sources across systems.

However it is done, there needs to be ready expertise with regard to the data and information models that are impacted, and they need to be documented so that other teams can understand the data that is broadcast to a data lake or that is provided via APIs and events. That means that whoever is maintaining the information models, data models, event models, and APIs, they must not be tucked away creating documents but are working with teams to help them to understand the data in an organization-wide or domain-wide context.

Agile 2 Principle: An organization's information model is strategic.

Agile 2 Principle: Create documentation to share and deepen understanding.

Agile 2 Principle: Both specialists and generalists are valuable.

Notes

1 itrevolution.com/the-three-ways-principles-underpinning-devops

2 www.slideshare.net/nicolefv/devops-a-value-proposition

3 hygeia-analytics.com/2018/09/05/new-science-explores-the-irrigation-paradox

4 itrevolution.com/the-three-ways-principles-underpinning-devops

5 steveblank.com/tag/customer-discovery

6 static.googleusercontent.com/media/research.google.com/en//pubs/archive/46593.pdf

7 testing.googleblog.com/2016/03/from-qa-to-engineering-productivity.html

8 Diana Larsen, Ainsley Nies. *Liftoff: Launching Agile Projects & Teams* (Kindle Locations 854-856). Onyx Neon Press. Kindle Edition.

9 Jez Humble, Joanne Molesky, Barry O'Reilly. *Lean Enterprise* (Kindle Locations 5114-5117). O'Reilly Media. Kindle Edition.

10 12factor.net

12

Agile 2 at Scale

When people say that Agile must work "at scale," what they mean is that it must work when applying it for many teams at once, to build or achieve "big things."

There is an argument that scale should not be needed: that one should always decompose any large endeavor into small independent endeavors—so-called independent autonomous teams—so that single-team methods can be used. However, as we have seen, it is difficult or impossible to achieve full team autonomy for complex products, even though autonomy and independent teams are powerful constructs and a worthwhile goal.

What about the product? Can it be separated into enough separate and completely independent subproducts so that each team can operate autonomously?

If you do that, then you do not have a product anymore. You have many small products. Imagine going to your bank's website and logging in to perform a function; but then when you want to perform a different function on the same website, you have to log in again to that function. No one would like that.

Or imagine the situation that Spotify had, where it had one set of teams for its mobile app and another set for its web app, and the two apps looked and behaved totally differently. That inconsistency was a contributor to why Spotify created its famous "Spotify model"[1]—to harmonize the work spanning the many different teams. From its website,

> "In the early days of Spotify, there was no design system—we were building everything for the first time. When we launched the mobile app in 2009, there were few standards or shared patterns in place, and the Spotify experience started to get increasingly . . .inconsistent. This drift continued until 2013, when we kicked off our first real attempt to align the visual design across platforms."[2]

But the challenges have escalated, and now in 2020 the different app versions have drifted again. See Figure 12.1.[3]

Feature	Mobile (Android)	Mobile (IOs)	PC App	WebPlayer
Edit playlist description	No	?	Yes	No
See playlist song count	No	?	Yes	Yes
See liked songs count	Yes	?	No	Yes
See friend activity	No	?	Yes	No
See personal top genres	Yes	Yes	No	Yes
Create playlist folders	No	No	Yes	Yes
Setup crossfade	Yes	?	Yes	No
Setup sound quality	Yes	Yes	Yes	No
Play folder as a whole	No	?	Yes	No
Change language in spotify itself	No	?	Yes	No
Support of Family Mix features	Yes	Yes	No	No
Copy songs across playlists	No	No	Yes	No
Add podcast to playlist	Yes	Yes	No	Yes
Swipe to queue	No	Yes	Obsolete	Obsolete
See new podcast releases	Yes	Yes	No	No
Hide songs in public playlists	Yes	?	No	No
Pop out album art	Obsolete	Obsolete	No	Yes
Sorting playlists by title etc.	Yes	Yes	Yes	No
Filtering playlists	Yes	Yes	Yes	No
Reorder playlists	No	No	Yes	Yes

Figure 12.1: Inconsistency of Spotify features across platforms

It is understandable why this has happened. According to Spotify,

> "In 2018, Spotify continued to grow, and fast. We had 200 designers. 2000 engineers. 45 different platforms. The days of designing for mobile and desktop were long gone. Now we were also designing for cars, smartwatches, speakers, and even smart fridges."

Today's products are complex, and getting more so. Consider a machinery product that contains many programmed control units, each communicating via a local network in the machine, with the machine sending diagnostic data to a gateway that feeds the data into a high-volume data stream, which posts the data to various stream processing systems to analyze the data, which in turn post analytics into various databases, which in turn are read by multiple product dashboard systems, depending on which the customer has subscribed to, and available on both web and mobile platforms.

There is no single team that can work across that stack; yet while the various functions can be decoupled to a large degree, they cannot be fully decoupled. A change to one can easily require changes to other functions on other parts of the stack, especially when new machine features are added.

The answer then is not to hope that all these things can be fully decoupled so that separate teams can be completely autonomous—they cannot. That does not mean that you should not try. For example, Spotify has systematized the way that it abstracts the various layers of its system so that a feature team can develop across multiple platforms. As we can see from the current situation, however, it has not entirely worked.

The reality is that while architectural abstraction is essential, one must figure out how to enable lots of teams that inevitably have interdependencies to maintain this complicated ecosystem in an Agile manner so that changes can be made quickly and with assurance.

Issues That Arise at Scale

Early Agile methods were not designed to work at scale. We are not aware of any evidence that most of the early Agile authors were thinking about scale or even beyond a single team. That is one reason why early attempts to use Agile in large organizations stumbled. For Agile methods to work in a large organization with a complex technology platform, one must have a development ecosystem that supports Agile teams that are mostly but not completely autonomous.

The challenges of scale are numerous. Some of the challenges for building complex products requiring many teams are as follows:

- Who creates the vision for the product, and how do they split their time among teams?
- Harmonization of how to decompose implementation of a feature that must span multiple teams, because no one team can implement the feature across the entire stack
- What architectural abstractions can reduce the dependencies between different products and different teams, how are those abstractions maintained and updated, and what happens when an abstraction needs to be violated to implement a new feature?

- Who participates in product-level decision-making, and how are decisions made?
- Who coordinates the many teams, and how?
- When a problem arises, who decides which teams need to be involved in addressing the problem?
- Which people or teams are responsible for the product-level outcomes, including technical outcomes?
- Who gets to talk to product designers and possibly actual users?
- How do groups of experts, such as experts in specialized technology such as machine learning or lab-based technologies that need to be "brought to market," interact with the teams, and with which teams or which individuals?
- If the organization has multiple products, how is funding allocated among them?
- How are product goals aligned with organization strategies?
- If there are multiple stakeholders, who coordinates and balances their needs?
- If there are risk-related stakeholders (e.g., required through laws and regulations), how are those stakeholders included?
- Do some teams need to work differently than other teams?
- If there is more than one product, how are the products coordinated in terms of behavior and technical integration with the technology platform?
- How does one manage the knowledge of people so that they grow and so that each team has the skills that it needs?
- How does the organization stay current?
- What shared services are needed to support all these teams and products, and what are the support models for them? How are those services funded, and how is their effectiveness and value assessed and continually improved?
- How is all of this continually monitored and refined as a system?

All of these scaling questions were answered long ago with traditional management models. We are saying that the traditional answers are hugely inadequate for organizations working in rapidly moving knowledge-intensive environments, and we need new answers—Agile answers—for these questions.

Various frameworks have arisen that attempt to address the scale problem. Some of these are the following:

- Scaled Agile Framework (SAFe)
- Large Scale Scrum (LeSS)
- Disciplined Agile (DA)
- LeadingAgile
- Nexus
- Scrum@Scale

There are good ideas in each of these frameworks, although some contain ideas or leanings that we do not think are helpful in many situations. So these frameworks are generally a mixed bag of good ideas and in some cases questionable ideas. That does not make them bad—not at all—and many of these frameworks contain great practices, and so they are all worth checking out. But it means that one should not apply any of these wholesale; to do so is to be careless, to fail to account for the uniqueness of your organization, and to implement something without actually understanding it and applying it contextually.

In other words, we feel that the biggest mistake is to implement one of these frameworks in its complete form, without tailoring it to the situation. However, there is nothing wrong with applying select ideas from some or all of these frameworks, if one does so thoughtfully.

Agile 2 Principle: Fit an Agile framework to your work, your culture, and your circumstances.

Agile 2 was conceived with the realities of Agile in mind, at scale, as we have experienced it in a wide range of organizations and situations.

What Is the Strategy?

All issues begin with establishing what the organization's goals and strategies are.

Some organizations place big bets. When Federal Express was launched, its big bet was that a private overnight delivery service would be widely accepted and could be profitable. Market research showed that there was a need for that, but the investors in FedEx made a big bet, because if it turned out that the need was not actually there, all of their investments would have been lost.

Organizations that undertake a transformation often make a big bet to reinvent themselves, either in the interest of survival or in a bold attempt to capture a new opportunity. SpaceX has a profitable satellite launch business, using the Falcon 9 and Falcon Heavy rockets of its own design and manufacture. However, its goal is not to make money; it is to establish a self-sustaining city on Mars, so the satellite business is only a step along that path. Right now it is making a big bet on an entirely new rocket design, which it calls Starship. Starship will replace both the Falcon 9 and Falcon Heavy in all of its missions.

If it turns out that Starship is not viable, SpaceX might well go out of business. But if it works as expected, SpaceX will be closer to its actual long-term goal.

Most organizations place smaller bets. When Apple launches a new type of watch, it is not betting the entire company. If the product fails in the market, Apple has the resources to absorb the loss, and it has many other products to sustain it. It is also worth pointing out that the small bets placed by an organization need to fit into an overall strategy, rather than being disconnected and random. Otherwise, they will not sum to a larger, more holistic bet.

A new feature set is a market bet as well. Apple's "butterfly" keyboard that Apple introduced on its Macs was a market failure, and in 2020, after numerous user complaints, Apple replaced it with a design almost exactly like the older "scissor" design.

Whatever the bet is—a company-wide bet, a product bet, or a feature bet—it is an experiment, and behind that experiment there is a hypothetical value proposition: the bet will pay off by advancing the organization closer to its goal. The bet is a strategy that tests a model of how the market will respond: "If we do this (the strategy), then the market will do that (the hoped for outcome)." If the bet is successful—the strategy works—the model is true; otherwise, it is false.

Strategy and Capability Alignment

Organizations often have many bets underway. They might also have ongoing business or activity that is based on prior successful bets. For those, the strategy is "keep doing what we are doing and make continual small bets to try to improve it."

These bets need metrics that help determine if the bet is working. In the book *Lean Analytics*, Alistair Croll and Benjamin Yoskovitz describe the concept of One Metric That Matters (OMTM) as a single metric that is most important that indicates whether you have reached your current goal, depending on what capability you are trying to build at that moment.

To tell if a strategy is working, it helps to have an OMTM. If the OMTM is "over the line," then the strategy is succeeding; otherwise, it has not yet succeeded.

That does not mean that the strategy was a complete failure. A failed strategy often provides important information that informs the next strategy. If eventually a strategy wins before one gives up or runs out of money, then the sequence of strategies has succeeded.

Each strategy requires capabilities. For example, if the strategy is to move one's storefront retail products online, then one capability that will be needed is an online retail presence. To implement the capability, one designs it at a high level and then attempts to decompose it into sub-capabilities that can be built and tested individually (when possible) or in sequence (when necessary). Each of these is an experiment and deserves its own OMTM for success, and each might need to pivot until success is achieved.

Note that a capability is not necessarily something functional. It might, for example, be "the ability to inspire user loyalty."

Each strategy needs leadership, and each capability needs leadership. These might perfectly align in some cases, but not always. There are also many dimensions to each: there is a business dimension and a technical dimension. There might also be a human resources dimension to provide the people who will stand up, operate, and maintain each capability. In Chapter 14, we will reframe the human resources dimension in a holistic manner.

If the organization has more than one product and the products interact, are used by the same user communities, or share functionality, then the products probably need to share components, or they might even share a common "platform" that provides a set of foundational functions for the products.

The platform needs leadership. Each product also needs leadership, but leadership that is not too invested in the product until the product has proven itself. Each feature also needs leadership, but not too invested in the feature until it has proven itself.

Portfolio and Capability Intersection

Large organizations often have a large footprint in their ongoing business or activity—the fruit of yesterday's bets—as well as multiple new strategies in play, yet they have aggregated sources of funds that need to be allocated to these strategies. Which existing activities or sources of revenue or investment should fund the new bets, and in what proportion?

Thus, for a large organization, there is not a single product vision; rather, there is an organization-level set of goals, being advanced through strategies, which require capabilities, some of which can be characterized as products to be developed. Each capability therefore needs a vision, or a concept for what to present to users, as an experiment to test a theory that those users will respond in a certain way.

The challenge is to balance the funding needs of the many ongoing activities and experiments that are underway. A particular challenge is that the capabilities often cross organizational boundaries.

Value-based management[4] and *value stream accounting*[5] are approaches that attempt to rationalize organizational spending based on value, instead of based on organization structure. These are beyond the scope of this book, but the technique of a Lean portfolio or portfolio Kanban model is based on the ideas of value-based management.

A Lean portfolio[6] as described by the Scaled Agile Framework is one model for managing a technology investment portfolio. In that model, a portfolio Kanban[7] is maintained. It is similar to an Agile feature backlog, except that each feature is an *epic*, which you can think of as a capability or a portion of a capability. Each epic has a vision, an MVP defined, and a Lean business case, which is a lightweight (single-page) expression of the model being tested by the epic. Epics are prioritized in the backlog based on a weighted shortest job first calculation, which attempts to balance value, effort, and the cost of delay.

That is an oversimplification, however, because there are often dependencies between epics, and some might have more certainty than others. All these factors need to be considered. Thus, treating them as a prioritized list is just a first approximation, and thoughtful discussion is still needed. Economic modeling is sometimes used as well to project future expected value.

An important thing to remember is that improvements to the systems for product development and delivery are capabilities as well, and if those are allowed to deteriorate

Agile 2 Principle: Don't fully commit capacity.

or become obsolete, then business capability delivery will suffer. Thus, some capacity needs to be reserved for ongoing improvement to technical processes.

Also, each epic should be outcome-based. An epic is not a project; it is a capability or a goal. The implementation teams must not be constrained by portfolio planners in how the epic is achieved. If one over-specifies the way an epic is implemented, then the implementation of the epic will not be able to proceed in an Agile manner, such that the product owner is able to pivot on individual product features.

The ability to pivot tactically is foundational for being Agile. If all future development capacity is committed to specific product features spanning a long duration funding cycle, then capability owners will not be able to redirect efforts as they learn from market-facing experiments or as circumstances change.

Need for Hierarchy

We have described what sounds like a hierarchy, which can be decomposed as follows:

> **Organization**: has one or more,
> > **Strategy**: has one or more,
> > > **Capability**: has one or more,
> > > > **Product**: has one or more,
> > > > > **Feature**.

When we say "has one or more," we mean that the hierarchical level contains one or more levels that are immediately subordinate to it.

This hierarchy is typical, not universal, but we will use this hierarchy for our discussion since it is very common. The hierarchy might not map to the organization's departmental hierarchy, which might be organized by market. Some capabilities might serve customers in multiple markets.

In addition, there might be platforms that support multiple products, and there might be features that span products. For example, a logon feature might be used by all of the organization's products, or a bank's customer balance system might be used by many digital products.

Products also usually consist of one or more components, and a feature might span many components. Some components might be used by many products.

Thus, instead of a single hierarchy, there are likely at least two overlapping hierarchies: the organization's departmental hierarchy, which might be market segment-based, and the capability and product hierarchy. Layered over that is a set of shared platforms and components, which is hierarchical but not strictly so because some components might be used by multiple products, and products might even embed other products. The interactions among these is also another topology. There is also a set of value streams, each of which consists of a path through capabilities that results in revenue or other value, that is, to organization goals being met and "bets" proving themselves; and value streams can intersect, converge, and diverge.

The situation is therefore not a simple hierarchy—not even close. Rather, it consists of a network of overlapping hierarchies and semi-hierarchies, as well as paths of value through the network—paths that need to be managed. (When we say "managed," we use the term in its pure sense—not in a traditional sense.) Choosing to manage only one of the hierarchies or paths will not suffice. They all need to be managed, even though they overlap; and the models to be used for "managing" these need to be agile and adaptable, rather than slow and rigid.

We hope that this overlapping of paths sounds familiar. In Chapter 5 we explained different kinds of leadership that are often needed, spanning multiple overlapping aspects—dimensions—of the organization. That is exactly what we are talking about here. Let's revisit the leadership issue.

Initiative Structure and Leadership

Each level in each hierarchy, each shared thing, and each path through this complex system needs leadership, and the leadership needs to change as the system changes, as bets prove themselves or fail or new bets are placed.

> **Agile 2 Principle:** Leadership models scale.

Decisions pertaining to each bet need to be made whenever circumstances change. Decision-making needs to be timely and have low overhead. It should not be ad hoc, however, because then it will be unreliable and things will be missed. That is why leadership is needed for each part of the system—not to dictate, but to orchestrate decision-making.

Let's consider an example. Suppose there is a company called Everywhere Doctor that provides telemedicine services to the public. Currently its services are available via the Internet, in which a patient uses video to communicate with a doctor. Patients who need to have a test must visit a testing lab, either operated by Everywhere Doctor or by another lab company,

> **Some of the insights that some members of the Agile 2 team had when discussing leadership at scale:**
>
> **Insight:** Teams of teams need leadership, just as teams need leadership. Each level of an organization is fundamentally a team.
>
> **Insight:** The various kinds of leadership that are needed at the team level are also needed at other levels, and vice versa. In the words of Peter Drucker, an outside person, an inside person, and a person of action.

and then go back home and wait for the result to appear in their Everywhere Doctor web portal, at which time they must schedule a follow-up video session with their doctor.

One challenge is that in remote areas, a patient might have to drive a large distance to a test lab—perhaps 100 miles or more—since labs are not generally found in remote communities.

A technologically savvy engineer on a software team mentioned to a business manager that there is new technology for performing many lab tests automatically, so-called PCR-on-a-chip (PCR stands for polymerase chain reaction), and that might make it feasible to create automated "kiosk" test labs that could be anywhere. The business manager realized immediately that would mean that remote communities could be better served. The question was, was it actually feasible? A second question was, would it be cost effective?

A team of company scientists and business analysts was asked to look into the use of PCR-on-a-chip technology and assess if it would be feasible to set up kiosk-like labs. They researched the

Insight: Points of accountability for product and technology collaboration are needed at every level, from the team level to the product level and outward.

Insight: Cross-functional issues need collaboration and issue resolution processes, at every level of an organization. The agility of those processes have an increasing impact on organization agility, because issues are becoming increasingly cross-functional.

Agile 2 Principle: Technical agility and business agility are inseparable: one cannot understand one without also understanding the other.

availability and maturity of the applicable technologies. In the course of that research, they identified a company that licenses technology for robotically drawing blood. The team reported that yes, the technologies needed were ready, the individual elements had received regulatory approval, and the team believed that it could be done in a kiosk setting.

The initial value proposition of the idea would be that a patient would not have to drive as far to have blood drawn. A second value proposition would be that if some tests could be run on the spot, using PCR-on-a-chip technology, that would allow the patient to receive the results during their visit and be able to schedule a follow-up session with their doctor right then.

The main question then became, would the return on investment be enough to justify use of scarce investment funds? Or should new investors be sought for this initiative? What were the risks? For example, could others simply copy the idea, thereby reducing the value of Everywhere Doctor's investment?

Currently no one had such a service. Everywhere Doctor realized that if it introduced this, other providers would copy it, but the company hypothesized that by being first, it had a chance at establishing itself as a reliable and trusted brand that was first to introduce this new way of performing medicine—its brand could literally become a household word. Everywhere Doctor also believed that if it got a foothold in this new model, it would be able to convince large footprint retail companies such as Walmart to introduce Everywhere Doctor's remote lab service to their stores, firmly establishing the service with a long-term business model and a stable revenue stream that would enable expansion of the service and investment in additional kinds of tests that can be run this way.

Financial analysis revealed that the investment would be worthwhile if it could lead to the establishment of a significant new brand for Everywhere Doctor, but if it would merely make life easier for rural patients, then the investment was too risky and could not be justified. That was disappointing to the team that came up with the idea, but they could not argue with the numbers: Everywhere Doctor was simply not large enough to take the risk unless the upside was extremely high. The team considered creating a subsidiary that could be capitalized separately, but the idea was rejected because the company already had an acquisition that it was overseeing and did not want to create another entity at that time.

The company surveyed the public and found that many people said they would use a kiosk blood test system. However, saying that they would use it did not mean they actually would. After all, who wants a robot to stick them with a needle? The company realized that the only way to see whether their idea would work was to try it on a small scale. The company also realized that its strategy of being first, leading to ubiquity through brand recognition and partnership with Walmart and others, could not be tested except by seeing it through. In other words, it was a "big bet."

There was a board meeting about the proposal, but the dilemma was that there was not enough data to prove that the strategy would work. The leadership team was in favor of going forward, but the board knew that often groups of people who agree about an idea can be wrong—a team cannot always be trusted, and a bad decision on this could break the company.

A conservative board would probably say no, but one member of the board had experience dealing with large retailers: she was previously able to launch a service that was installed into every Costco. Based on her experience and the particulars of this situation, she convinced the board that they should proceed with the proposal. They trusted her, because she had been right many times before—she was a proven visionary—and her explanation was clear and convincing.

One of the board members, Fred, had had a bad experience trusting a "visionary." He learned that visionaries can be wrong and that often it is better to rely on careful market analysis. There was a fear, however, that the window of opportunity would close. The PCR-on-a-chip technology was new, and it was only a matter of time before someone else commercialized it, established themselves as first, and built early lucrative relationships with large retailers, so Fred decided to accept the risk and did not challenge the proposal as long as they took a gradual approach and test-marketed the idea.

To test the viability of a remote lab, the company developed a basic version of the lab and set one up in a grocery store in a rural community where Everywhere Doctor already had telemedicine patients and did not have a conventional test lab. The company

Agile 2 Principle: Validate ideas through small, contained experiments.

advertised the new service heavily to its patients in that community. They began by offering only basic blood drawing service in the lab.

This initial lab kiosk was what they considered to be a minimum viable product (MVP). A robot performed the procedure, just as a human clinician would, using a setup much like the automated blood pressure tests that one finds in supermarkets. The sample was then refrigerated and held for pickup by a courier. The company called the service Everywhere Doctor *FamilyLab*. Thus, this MVP did not actually use the PCR-on-a-chip technology that was the impetus for the idea in the first place. The MVP's purpose was to see if people would actually be willing to use a lab kiosk that takes their blood; the value to them was that they did not have to drive to a faraway testing lab.

The MVP system therefore consisted of a robot that was able to reliably draw blood from humans, as well as software to schedule and coordinate the use of the remote lab by patients. There was also a courier service to pick up the samples that were collected. The ability to analyze them on site was left for a future version of the service.

The company's key business strategy was to establish a novel marquee brand service and leverage it to place its service into large chain stores such as Walmart. The capability proposed to achieve that was a service consisting of a fully automated blood test lab called *FamilyLab*. The strategy and capability breakdown was as

Agile 2 Principle: Any successful initiative requires both a vision or goal and a flexible, steerable, outcome-oriented plan.

follows. (Please do not focus on the terminology, such as *capability* versus *product* versus *feature*, as that is not important and is somewhat arbitrary. What is important is the hierarchical decomposition.)

Company: Everywhere Doctor; has one or more,
> **Strategy**: to be first to create brand recognition and strategic partnerships; has one or more,
>> **Capability**: the FamilyLab service; has one or more,
>>> **Product or sub-capability**: Lab kiosk; has one or more,
>>>> **K.1 feature**: Secure system that can seat a patient, verify their identity, securely record the details of their visit, and terminate the session when the patient exits the equipment.

K.2 feature: Robot that can draw blood and place the sample into a receptacle, such as a refrigerator.

Product or sub-capability: Blood test scheduling system; has one or more,

S.1 feature: Ability for a patient to schedule access to a local FamilyLab kiosk via Everywhere Doctor's current telemedicine portal.

Product or sub-capability: Result entry; has one or more,

R.1 feature: Automated entry of lab results into the patient's online Everywhere Doctor medical chart, and notification of the patient and doctor that results are available.

Needed *only* for MVP:

Product or sub-capability: Courier; has one or more,

C.1 feature: Daily pickup of refrigerated samples and transport to a regional FamilyLab test lab where actual blood tests are performed and results logged.

This sub-capability is only needed for the MVP service.

Needed for full service:

Product or sub-capability: Blood test panel; has one or more,

P.1 feature: Automated blood test using PCR on-a-chip. (The range of test types can be expanded over time.)

The company realized that the set of capabilities that would be needed long term were somewhat vague. What kinds of tests needed to be supported? And how important was the instant turnaround of results for each of those? These were refinements if the service could be established and proven viable.

To make this happen, leadership needed to design the organization to support this initiative. They realized immediately that this initiative had significant growth potential, and they needed to establish a robust leadership paradigm and culture from the beginning. Since the service was highly technical, they also realized

Agile 2 Principle:
Product development is mostly a learning journey—not merely an "implementation."

that issues would be complex, and so decisions that arose would not be easy: they needed ideas to be discussed in a very thoughtful way, and all minds would be needed.

The Everywhere Doctor leadership team decided to create a FamilyLab leadership team consisting of the following:

> **Agile 2 Principle:** Both specialists and generalists are valuable.

- A doctor: Lauren
- A former test lab director: Jose
- A roboticist: Andy
- A manager who was a proven organizer and servant leader: Tamika
- A regulatory lawyer, as a part-time "extended team member": Rick

They made Tamika the team lead, meaning that Tamika would organize and facilitate the group's activities, but also empowered her to make final decisions if she felt she needed to. Large issues were expected to be escalated to the Everywhere Doctor leadership team. The initiative was given a funding stream—a monthly budget.

The Everywhere Doctor leadership team then worked with the FamilyLab leadership team to establish baseline performance and financial models for the FamilyLab service. These models included objectives and key metrics. A dashboard was sketched out, and an analyst was assigned the task of creating and maintaining the dashboard—and automating parts of it where possible. The objectives were based on the long-term goals and anticipated milestones along the way, establishing the following:

- People will use the kiosks.
- The robot technology/process is reliable and sanitary, and it complies with all applicable government food and drug regulations.
- The system is secure and complies with all healthcare laws and other applicable security requirements.
- The PCR-on-a-chip technology can be deployed and is effective in a kiosk setting.
- The value proposition to patients of not having to drive far is realized.
- The value proposition to patients of getting immediate results is realized.

Metrics were proposed for each of these, but the only ones that were nailed down at the outset were the first three.

Tamika then emailed her leadership team and wrote, "Let's get together tomorrow for an all-day session and map out how we are going to do this."

Agile 2 Principle: Good leaders are open.

Jose thought, "Perfect. That will give me time to collect my thoughts in private before we meet." He spent a couple of hours organizing his thoughts and then sent a message to the group, outlining what he thought they should do. Andy responded in detail with some questions and challenges, but no one else did.

The next day the team met. Tamika brought an analyst with her and introduced the analyst to the team: "This is Joseph. He will be supporting us as an analyst." Everyone welcomed Joseph.

Tamika then said, "I read the emails from Jose and Andy. Has everyone else?"

Rick said, "I skimmed it—it was kind of long." Tamika thought, "He didn't really read it." Lauren said, "I read the emails. I thought they were a good start."

Tamika made a mental note to watch out for Rick not staying abreast of written discussion, but reminded herself that he is a part-time team member, and she might have to reach out to him to make sure he is current on important issues that arise.

Agile 2 Principle: Foster diversity of communication and diversity of working style.

Tamika then drew on a whiteboard, describing what Jose and Andy discussed and listing their current point of disagreement. She then asked, "What do others think?"

Throughout the discussion Tamika kept a list on the board of the concerns that were raised. They talked through each and eventually reached consensus on an approach for how they would build, deploy, and operationalize the MVP. Joe agreed to document their approach and put it online with their dashboard.

The approach that they outlined was as follows:

The following teams will be created:

- Team A
- Team B
- Team X

Teams A and B will have members with these skills:

- Robotics
- Internal medicine
- Hardware/software system engineering
- Software engineering
- Hardware and software test automation
- Organizational skill and demonstrated servant leadership

Teams A and B will each have these *extended* (part time) team members:

- Engineer on call from the blood drawing robotics company
- Rick, for legal counsel, as needed

Team X will have members with these skills:

- Product design and user experience
- Patient customer service
- Organizational skill and demonstrated servant leadership

Each team's organizing member will be designated as that team's lead. The lead will have these responsibilities:

- Be able to speak for the team.
- Be responsible for ensuring that collaborative decisions are made among the team.
- Be the final arbiter for team decisions; but not its "dictator."

- Be responsible for ensuring that team metrics are always up-to-date and accurate.
- Be responsible for resolving team problems of any kind.
- Be responsible for creating a positive and inspired atmosphere among the team.
- Ensure that knowledge and skills needed by the team are obtained.

Jose said, "If this takes off, these teams will become permanent teams. In that case, what about HR type issues? We'll need to start thinking about professional development and career issues of team members."

Agile 2 Principle: Professional development of individuals is essential.

Tamika said, "That is a great point. Let's add that now. Let's add that the team lead needs to also be responsible for ensuring that team members have professional development and leadership development." Everyone nodded in agreement.

Jose said, "Who creates these teams?"

Tamika said, "How should we handle that?" and looked at the group.

Jose responded. "We'll need to collectively recruit them, because they are cross-functional."

Andy said, "Yes. Some can come from Everywhere, but we'll need to bring in some new people."

Tamika then asked, "Who will take on a leadership role for each feature?"

She got blank stares, and then said, "The features cut across teams. In my experience, if someone is not on point for something, it suffers. When I say 'take on leadership for a feature,' I mean that you are that feature's advocate, shepherding it and making sure it gets done right. You are not managing the feature—you are looking out for it. And you will be expected to measure its progress and quality in some reasonable way."

Lauren said, "I can lead the blood test drawing feature."

The others volunteered as well, and the result was this table:

Feature	Who Leads
K.1 feature: Secure system that can seat a patient, verify their identity, securely record the details of their visit, and terminate the session when the patient exits the equipment.	Jose
K.2 feature: Robot that can draw blood and place the sample into a receptacle, such as a refrigerator.	Lauren
S.1 feature: Ability for a patient to schedule access to a local FamilyLab kiosk via Everywhere Doctor's current telemedicine portal.	Andy
R.1 feature: Automated entry of lab results into the patient's online Everywhere Doctor medical chart, and notification of the patient and doctor that results are available.	Andy
C.1 feature: Daily pickup of refrigerated samples and transport to a regional FamilyLab test lab where actual blood tests are performed and results logged.	Tamika

Lauren said, "So we're good for now."

Tamika looked at the clock. They had been in the room for three hours. She said, "No, we're not" and smiled. She looked around the room and then back at Lauren and said, "Lauren, your feature is arguably the most critical one for the MVP. How do you plan to lead it?"

Lauren said, "At Everywhere Doctor I had been advising the Care Delivery group, which is under Robin Sobel, but before that I was head of internal medicine at a Chicago hospital. We had Lean dashboards, and I used to manage productivity."

Tamika nodded and said, "This is pretty different, so I'd like to help, because I have some ideas. I'd like to work with each of you on how you approach this, because each of you might need to use your own approach. These teams have their own leads, so you can't tell them what to do. And we are going to have a prioritization process for their work, so you can't mess with priorities either. But what you *can* do is troubleshoot and watch for things that are being missed."

There was silence, and after a few moments Andy said, "So, we're like expediters."

"Kind of," Tamika said. "But if something is not going the way you think it should, you need to be able to get people's attention. I think that the feature leads need to be explicitly designated as such so that you have standing. You can call a meeting. You can say, 'What about this?'— what do you all think? What does 'leading a feature' mean to you?"

Jose said, "Yes, I think I get where you are going. You know that I ran one of our test labs and had to hand that off to do this. At the lab I had different test process leads, and I learned to not micromanage them. I used to ask powerful questions, and we had performance metrics."

> **Agile 2 Principle:** The most impactful success factor is the leadership paradigm that the organization exhibits and incentivizes.

Tamika said, "Yes. And did you also have experts or technical leads who were in charge of certain machines or steps that were used by multiple test lines?"

"Yes. At Everywhere Doctor we call them Med Tech Leads. They advise the lines on how to use the machines."

Tamika continued. "That's thought leadership. They are not in charge of people, but they can set standards and advise the lines."

"Yes."

Tamika: "And what about each test process? Were there people in charge of a test line?"

"Yes. But the good ones don't tell everyone how to do their job; they act as coordinators and organizers."

> **Agile 2 Principle:** A team often needs more than one leader, each of a different kind.

Tamika again: "And when there is a new technology or machine introduced, how does that work?"

"We pull people from the teams who pilot a new process. We also look at the skills needed by each line and see what training people will need. And when we stand up a new process that we have piloted, we have ongoing feedback on how well it is working. We have quality and process time metrics."

Tamika said, "Okay, I think we are all spent! Is there anything else we need to work through today?"

Andy: "We did not actually say how we are going to get started."

Tamika said, "Right. How about we all mentally process this on our own. Then tomorrow meet again for an hour and come out with actions?"

Everyone nodded in agreement, and they ended the meeting.

How do *you*—the reader—think they did?

They not only defined the organization, but they defined important dimensions of leadership. While doing that, they specified leadership styles that they felt are important for the leadership roles. They also set some expectations about measurement, communication, and accountability. Interestingly, none of them came away being in charge of anyone. Instead, each volunteered to lead one or more features, and they defined what they meant by *lead* in that context.

They were not done. They were only at the beginning of their journey for this MVP; but they had defined a framework containing the most important things: the things that established the guardrails and the cultural norms that would be the foundation of this new service, FamilyLab.

Notice that this team of leaders was also just a team. Tamika was not told to be in charge of the group. She naturally assumed that role, as a facilitator, but did not tell anyone

Agile 2 Principle: Leadership models scale.

what to do. Instead, she used a Socratic process, asking questions and sometimes making suggestions in an inquisitive way, making it clear that she was not dictating. She also offered to work with some of the team members to help them to think through how they would approach leading their feature.

The word *Agile* never came up, but it was present in the things that were discussed and the approaches that were used. No Agile framework was proposed either—it was not necessary. The team defined their own framework, appropriate for their situation.

Coordination at Scale

Fast-forward two years. The FamilyLab initiative is now a profitable Everywhere Doctor service line, and the kiosk labs now support automated PCR analysis on the spot with results available in 30 minutes. Everywhere Doctor has partnered with Walmart, and the FamilyLab service is now installed in more than 40 Walmarts, with more planned, and expansion to other retailers in several countries is being pursued.

The FamilyLab service has continued to evolve into what was needed at each stage, because it conducts recurring "retrospective" discussions about its methods. Those discussions occur at each level: each team and the whole ser-

> **Agile 2 Principle: Organizational models for structure and leadership should evolve.**

vice. The discussions include face-to-face discussions (in-person or online) so that those who are most energetic in conversation could express their ideas, but a discussion forum is also created for each issue that is identified. That has enabled the quiet thinkers to chime in too.

There are now eight product development teams, A through H, and the X team (discussed later in this chapter) has grown to six members, three of whom are collocated and the other three remote,

> **Agile 2 Principle:** From time to time, reflect, and then enact change.

but all within the continental United States.

There is another new team, the Product Management and Operations team, aka the PMO team, that owns the product vision, the medical efficacy and operational effectiveness of the FamilyLab services, and FamilyLab's profitability. Of the original FamilyLab leadership team, only Tamika is full time in the PMO team; the other members are "virtual" members, as they have moved to lead other teams.

In fact, except for Tamika, the PMO team is entirely a virtual team, in that all of its members except for Tamika belong to other teams. The intention is to empower the other teams but aggregate their

> **Agile 2 Principle:** Foster collaboration between teams through shared objectives.

influence and leadership around the issues that determine overall product success.

The PMO team has final decision-making authority about product vision, high-level capability definition and prioritization, and deciding which new capabilities are market-ready. They do not define or prioritize product features; the X team still does that.

There is also a set of R&D teams that are continuously working on new types of tests that can be automated. Those teams are located in a research lab facility in Boston.

Five of the development teams are entirely remote. Three of them have members in Bratislava, Prague, and Copenhagen. Another team has members in Cupertino, San Jose, and Portland. The fifth remote team has members in Chennai and Mumbai.

Each development team now not only has an "organizer" who serves as a team lead, but also has a technical lead who facilitates discussions about cross-functional technical issues. The technical lead is also empowered to arbitrate technical decisions when necessary, but consensus is strongly preferred. The technical lead role was decided upon by the team coordinators to solve several problems:

- The FamilyLab initiative had grown, and there were too many people across the teams to collaborate on every technical issue, so a technical representative was needed from each.
- Teams often overlooked cross-functional issues. It was felt that someone needed to be always looking "end to end" on the technical issues, because of their complexity.

The technical leads of the various teams are not able to meet in person due to their geography, and it is difficult for them to have a weekly face-to-face because of their time zones, but despite the time zones, they have a video call that they all join once a month. On a daily basis, they communicate through a Microsoft Teams channel and maintain an online list of issues that they are working through. If someone observes an issue that needs to be discussed, they add it to the list and may also start a Teams thread for the issue.

One issue that arises frequently is that the work of one team is dependent on the work of another team. For example, the R&D teams have proven a new test, and the test needs to be rolled out to the kiosks. To do that, features need to be added to the lab kiosk, the test scheduling system, and the result entry system, which is now called the result entry *gateway*. Deployment to the many kiosks also needs to be designed, and it will include physical installation of equipment, over-the-air embedded software updates to equipment that is in place, and updates to server software.

When the MVP was turned into a real product, a good decision was made. An end-to-end mock lab was created to test deployments and test the operation of the kiosk equipment. The mock lab is outfitted with the same equipment that the real field kiosks are.

Agile 2 Principle: Integrate early and often.

In addition, within the mock lab, whenever a piece of equipment is replaced, the obsolete one is left in place. Instead of replacing it, the new replacement equipment is added. The equipment is software-selectable so that automated processes can choose to use either the old equipment or the new replacement equipment. As a result, it is possible to run automated tests with either. This has proven valuable, because as equipment upgrades are rolled out, FamilyLab can test how software improvements work with both the old and the new equipment. It also means that as software is developed, the teams can run tests with either the old or the new equipment.

Having done this, it became possible to install multiple instances of each piece of equipment, creating a kind of equipment "farm," much like mobile device farms that are used for software testing on mobile device handsets. The result is that all the teams can run tests in parallel, without any contention for the mock lab's equipment. The tests simply provision the devices that they need from the pool of those that are available. This greatly improved the throughput of the development teams and made it possible to completely develop and validate new software features in days instead of weeks.

Another result of this innovation was that teams never have to wait for each other. If one team is developing a feature needed by another, the teams can decide on a set of integration tests that validate whether their changes work together. Thus, if the integration tests pass, it means that the work of each team is in sync. No team ever has to wait for an integration "phase" in which they converge their work. Convergence can happen all along. This was also a contributor to the very fast turnaround time for new features.

These improvements came about because the technical leads considered the problem of dependencies between teams, using an end-to-end "system" perspective, and realized that the dependencies were potentially a bottleneck. To remove that bottleneck, they needed to make it possible for teams to *never wait* for equipment or for each other. It was their thought leadership around the end-to-end issues that improved the overall delivery, spanning many teams. No one team could have solved these problems by itself.

The technical leads also struggled with the problem of how to ensure quality. Kiosk tests absolutely need to be as accurate as conventional lab tests, and there can be no mix-up with test results. They therefore discussed how to create a very high standard for quality, but without slowing teams down.

Their solution was to use automation, but to manage it. All tests must be automated, but a quality assurance (QA) team was set up, consisting of one member of each development team, to define criteria for deciding whether the automated tests are sufficient. The QA team defined a process in which whenever a new feature is designed, a set of test specs must also be specified. Those tests are then reviewed by each member of the QA team, who can add supplemental test specs. Thus, the set of test specs is enriched by the knowledge and perspective of each member of the QA team.

In addition, whenever a problem occurs in the field, pertaining to quality, the event is discussed by the QA team, and analyzed from the two perspectives of how the design process could have avoided that problem and how the test process could have caught the problem. Both processes are then enhanced accordingly, often by adding additional tests to the test suites and adding criteria to the system design.

What about product design? There is still a single X team, but it has grown to six people. They are distributed but within a few time zones of each other, so they are able to have video discussions. This is helpful since product design and usability are highly collaborative and creative processes. The X team lead and the team's product design lead both are members of the PMO team.

The X team operates in parallel with the development teams. Their job is threefold:

- Take feature ideas that the labs have validated and that are ready for productization and design the feature—that is, how it should actually work, both from a patient's perspective and from a "back room" lab processing perspective.
- Talk to patients to discover ways to improve FamilyLab, and propose new features or feature changes accordingly.
- Consider ways to improve the overall efficiency and effectiveness of the FamilyLab system.

As such, their team now contains these skills:

- Product design and user experience
- Patient customer service
- Organizational skill and demonstrated servant leadership
- Diagnostic expertise
- Biomedical engineering
- Operations research

They have also become the main driver for product features, based on the capabilities that the PMO team decides upon (two X team members belong to the PMO team). The FamilyLab PMO team works with the X team as well as the development teams and the labs to identify and define important new capabilities for the FamilyLab service. It also defines the value streams that are measured. Strategies spanning the various functions are discussed holistically among the PMO team. Jose runs the labs researching new test methods, and Andy runs the labs researching applications of robotics.

Since FamilyLab went live as a full Everywhere Doctor service, the X team has tried several approaches for organizing their work. Some of these were as follows:

- Participative design, by inviting real patients to come up with FamilyLab features
- Embedding in the product development teams, so that features are refined during development through collaboration with developers. The product designers still belonged secondarily to

the X team, but it operated as a "chapter" in which they could discuss cross-team issues.

- Flowing work to the product development teams, so that the teams receive fully refined product functional and visual designs
- Having a two-stage process in which a feature's design is fleshed out through collaboration between the X team and the tech leads of the development teams; and then the feature design is flowed to the development teams

Currently they are using the last approach, as well as the first approach in semi-annual customer workshops.

The PMO team is responsible for FamilyLab as a business, but they are not in charge of the other teams, although the PMO team does oversee those teams as well as reorganize them when necessary.

Agile 2 Principle: Organizational models for structure and leadership should evolve.

The development teams are now under a development lead, Samantha, who had been a team tech lead. It was felt that the person running the development teams needed to be a developer—someone who

Agile 2 Principle: Provide leadership who can both empower individuals and teams, and set direction.

understands that cohort of people. The development lead chosen was found to be a natural organizer and servant leader, who makes sure that issues get identified and get resolved and that everyone has a voice.

At one point the question arose of whether to merge the X team members into the development teams. Some team members had been on teams in the past that did that and felt that it worked very well. In the end, FamilyLab chose to keep X as a separate team to ensure a consistent and strategy-driven user-centric approach for the product.

In general, the development teams are allowed to decide how they work. However, there needs to be some commonality between them, or it would be difficult to coordinate how cross-team issues are dealt with. This is particularly true since all of the teams except for the "platform" teams (we will explain those shortly) operate mostly as feature teams, so they work on the same components.

They are not entirely feature teams, though. Since the FamilyLab system includes both hardware and software, some teams have imperfectly balanced expertise toward one or the other. For example, the C team currently has three Erlang programmers, who are able to work on the embedded software that controls the robots. Other teams have only one or two Erlang programmers, and Erlang is a difficult language, so it will be awhile before other team members are able to learn it. As a result, team C tends to focus more on robotic aspects of a feature, in partnership with other teams who complete the nonrobotic aspects.

That means that team C works a little differently from the others. There were a lot of conversations about that, and Samantha proved valuable for moving those discussions along and being able to understand the end-to-end issues herself but still encouraging the teams to reach a consensus. It was felt that Samantha would be effective at being sensitive to each team's need to use its own approach while also being aware of the things that need to be harmonized across the teams.

Observe that for FamilyLab there are now several overlapping lines of authority. There is the development lead (Samantha), there are the value streams, and there are the features. There is also now a core "platform," which we will describe later. And there are oversight functions that cut across all areas for legal compliance and other areas of risk.

This means that these different dimensions need to be reconciled in real time, through collaboration. For example, a feature lead, say Lauren, might feel that a feature is not progressing due to some issue and that the legal group needs to be involved in resolving the issue. At the same time, it could be that the feature has a low priority as set by the X team, and as a result the development manager has not found it necessary to get teams trained on some new technology needed for the feature.

Agile 2 Principle: Favor mostly autonomous end-to-end delivery streams whose teams have authority to act.

One way to reduce these intersections between the work of teams is to create development capabilities that automatically align teams along the various issues. For example, if there is a concern within the legal team that the development teams will not adhere to compliance rules, a compliance test suite can be created. From that point on, as long as the test suite is complete enough and is maintained (two important considerations), the development teams can verify their compliance merely by running the test suite. There is no need to collaborate with the legal team about compliance, because the compliance rules have been reduced to automated checks.

That is a "self-service" approach, which is a powerful technique. If the organization can create self-service capabilities that enable teams to automatically check their alignment, then teams can become autonomous about those issues.

Many cross-team issues cannot be automated, and issues can sometimes be complex and tangled, like the one described earlier. Resolving complex cross-team issues will require leadership—there is no way around it. The various stakeholders can start a discussion about it. In an organization, discussions can easily become political, because a decision affects resources and priorities.

> **Agile 2 Principle:** Foster collaboration between teams through shared objectives.

One source of politics is territory, as in my team versus your team or "that is my feature" or "that is my value stream." If people are measured based on "their" thing, they will naturally tend to protect that thing to the exclusion of other people's things. Thus, do not measure them that way. Make the assessment of individuals based on *how well they perform as members of the organization*, as well as on their shared outcomes; and to do that, you have to be close to the action. Otherwise, you will know only what you are told by those who are waging politics—you will not have seen things for yourself.

R&D Insertion

The services offered as part of FamilyLab are cutting edge and therefore highly dependent on the productization of new technology from the lab. That requires these two steps:

1. Validate the new technology as "ready" for use with patients in every sense.
2. Convert the highly manual and often inefficient laboratory-based process into a fully automated and efficient system that can be installed in FamilyLab kiosks.

The first step includes obtaining regulatory approval for each location in which the new test will be installed; the applicable laws and regulations depend on the locale of the kiosk. The first step also includes validating that the technology is technically ready, that it is reliable, robust, and fully automatable. The first step is therefore a gate before the second step can be attempted.

The second step requires collaboration between the lab researchers and the development teams. In most cases, a proof of concept is developed, and then its efficacy is assessed. Several attempts are usually needed, each improved on the one before. Each is an experiment.

Agile 2 Principle: Both specialists and generalists are valuable.

Collaboration between the lab researchers and the development teams requires leadership who understand both ways of working. Laboratory researchers are highly specialized and are focused on the technology in question and are trained to be methodical and extremely careful. In contrast, product developers are inherently more focused on automation and are trained to be fast and efficient. Bridging these two requires sensitivity and knowledge of both domains.

Multiple Stakeholders

FamilyLab's virtual PMO team has become the main driver for the FamilyLab service. Thus, the team works with marketing, interfaces with customers, and defines features, and now the team also sets the vision and strategy for the service and oversees efficacy, profitability, and overall business performance. However, there are other teams beyond those described so far, and each of these has a member in the PMO team. Some of the other teams are as follows:

- **Everywhere Doctor Global Legal and Compliance Team**: Ensures that all Everywhere Doctor products and services comply with all laws and regulations in each geographic region and performs reporting that is required by drug and healthcare laws.
- **Everywhere Doctor Global PeopleOps Team**: Ensures that Everywhere Doctor's people are developed and have opportunities to grow and also that all laws and regulations pertaining

to employee and contractor personnel are adhered to in each applicable geographic region.

- **FamilyLab Platforms**: FamilyLab now has some standard robot systems and software systems that are used by most features of the FamilyLab service. These are now maintained by a select group of the development teams to ensure that their design retains "cohesion" over time—in other words, that they do not become a patchwork of small and inconsistent changes, aka hairball.
- **Financial Operations**: Accounting and profitability analysis
- **Kiosk maintenance**: The people who go out and install, maintain, and upgrade FamilyLab kiosks
- **Walmart (partner)**: Since FamilyLab kiosks are now inside Walmart stores, Walmart has a say about those kiosks and must be included in the pre-rollout testing of any new feature or significant change.

The Global Legal and Compliance Team, which we will just call the *compliance team*, operates in an Agile manner, in that the team embeds a member part time in each Agile team. This compliance team member stays abreast of team discussions and watches for issues that might arise that might affect compliance. If they notice such an issue, they immediately raise it.

In addition, the compliance team members assigned to FamilyLab are expected to add test scenarios for the automated tests that are run for FamilyLab. As such, those members act as their own team and maintain the compliance portion of the automated test suite. The test suite may contain manual tests, but they are strongly encouraged to try to define every test as an automated test. The guidance is that if a new version of FamilyLab passes the tests for a given region, then it is safe to deploy to that region.

One concern was that the supporting teams would become sources of delay or resistance when change is needed. For example, will the Financial Operations team be cooperative if the PMO team wants to automate more of how financial data is collected and reported?

To address this, the charter of each supporting team states clearly that the PMO team is their "customer," and performance of each team is assessed continuously using agreed-upon metrics. That provides structure for dealing with the potential for conflict, but FamilyLab also addressed it on a human level, by including the lead of each supporting team as extended members of the PMO team, and having frequent collaboration and retrospective sessions with them, during which collective goals for improvement (and measurement) are developed.

FamilyLab discovered through market analysis that their customers can be divided into these overlapping categories:

- Elderly
- Rural
- GenX
- Nordic
- British

Each demographic has some characteristics and preferences that seem to apply across the other demographic categories. This means that feature testing now needs to be inclusive of these different demographic groups, and it means that the FamilyLab system now needs to be tailored to some markets—so there is no longer "one" FamilyLab system, but several options that are enabled depending on where a kiosk is located. This had already become necessary due to health laws and rules in different regions. Thus, there is no longer one product, but a differentiated product that is becoming more complex every day. FamilyLab realizes that managing this will be its next challenge and will require new approaches for how it is organized.

Knowledge Gap

One of the biggest challenges that FamilyLab has is that FamilyLab is a very complex system, involving state of the art diagnostic equipment, robotics, embedded software that is now written in Erlang, which is a language that most programmers do not know, and application software that runs on servers. There is now also data analytics that have been added, using data streaming systems including Apache Spark, as well as machine learning.

Few programmers are able to work with all of those tools, which is why FamilyLab product teams contain members who have

> **Agile 2 Principle:** Professional development of individuals is essential.

a range of skills. Most members are able to work on several but not all of the different aspects of a feature.

Knowledge is therefore a major constraint in who can work on what. To address this, FamilyLab leadership works with the Everywhere Doctor PeopleOps function to help team members plan for their professional development. FamilyLab also makes use of practices such as pairing to help team members learn new skills. They bring in experts to train and coach team members, and they sometimes request the research labs to provide training on new technologies.

Reflection on FamilyLab

We hope you have enjoyed the FamilyLab example. Please remember that it is *only* that: an example. It is not intended to be a template for how to do things.

In this book we have strenuously pointed out again and again that one should use Agile ideas thoughtfully and not copy what someone else has done. This applies to the FamilyLab example as well. We hope you can see that each of the decisions that FamilyLab made about how it does its work was influenced by the specific circumstances.

Consider, for example, FamilyLab's X team. We saw that over time, the X team tried several approaches for organizing their work. Each approach they tried made sense at the time, but eventually things changed, and another approach was needed.

We can also see that the issues that arose are not addressed by an Agile framework, and the main challenges did not pertain to how each team operates or whether it uses Scrum. Rather, the challenges pertained to integration across the many kinds of issues and teams, and what collaborative and leadership structures and practices were set up to manage those issues.

Finally, we saw the issues changed over time. Organizations are dynamic. Circumstances are dynamic. The way that your product is organized and operates today might need to be very different from how it was organized and operated a year ago.

There were many Agile 2 principles that applied to the various situations in the example, and we identified them along the way. We hope that you do not treat those principles as inviolate rules, but instead view them in the context of ways of looking at a situation to try to understand it and thereby come up with approaches to try.

Notes

1. blog.crisp.se/wp-content/uploads/2012/11/SpotifyScaling.pdf
2. spotify.design/article/reimagining-design-systems-at-spotify
3. imgur.com/YNlyg7a
4. www.mckinsey.com/business-functions/strategy-and-corporate-finance/our-insights/what-is-value-based-management
5. enfocussolutions.com/value-stream-accounting
6. www.scaledagileframework.com/lean-portfolio-management
7. www.scaledagileframework.com/portfolio-kanban

13 System Engineering and Agile 2

Today's products are often a combination of hardware and software. A car contains millions of lines of software code, as well as thousands of mechanical parts. An iPhone or Android phone contains an extremely complex operating system, but also a myriad of extremely carefully engineered parts. There is a lot of software in the world, but while many products today are exclusively software, most modern products are *not* software; they are physical objects that contain parts and often "embedded" software as well.

Some products are complex and contain other products, or *assemblies*. An assembly is a reusable component that consists of many other components. For example, an engine is an assembly, because it is a component in its own right, can be used in many kinds of car, and consists of many smaller parts. System engineering, aka *systems engineering*, is the discipline of designing complex hardware products, which may or may not contain software.

In the early days of computing, the process used for creating software was much like creating hardware. Software was created by laboriously crafting it and testing. This was not unlike a blacksmith forging a new tool, by hammering on it, inspecting it, hammering again, inspecting again, until finally it meets the blacksmith's expectations and the tool is declared done and placed in water to cool it. To inspect software, one runs it to verify that it produces correct output.

What about a hardware machine that contains many forged pieces, such as an 18th century wagon? Each piece was created individually, and instead of forging, other techniques such as milling might be used, because they are faster and more repeatable. These pieces could be created ahead of time and added to an inventory, but still each wagon was created individually and in entirely. This method continued when wagons began to be replaced by powered vehicles—cars and trucks.

Then in 1901 Ransom Olds introduced the assembly line concept,[1] which created a pipeline that produced vehicles one after another, by assembling premade parts, in a highly repeatable manner. The assembly line method was a continuous delivery method, which today is included under the umbrella of DevOps.

The assembly line is limited, though, in that it can only make many replicas. An assembly line cannot make unique products.

The idea can be adapted, though, as we explained in Chapter 9. If one looks at an assembly line as a pipeline in which each step can be augmented over time, so that better and better parts are added and more parts are added, and the line can be reconfigured as the product becomes more complex, then we have a true DevOps pipeline that can create a continuous stream of ever-better versions of the product.

A pipeline is a product in its own right that makes our actual product—the one we want to sell. To improve the product over time, we improve the pipeline and its inputs over time.

Continuous delivery can be used for hardware. To use continuous delivery for hardware, we need to create a pipeline that makes the hardware. The pipeline produces a continuous stream of copies of the product. To develop and refine a product over time, we develop and enhance the pipeline.

Agile 2 Principle: Work iteratively in small batches.

How Hardware and Software Differ (or Not)

There are important differences between hardware and software. For one, copying software costs almost nothing, whereas to copy hardware, one needs physical material, and its cost might be substantial. Each time a hardware pipeline produces a copy of the product, the pipeline consumes many components and raw material, and these each have a cost. Therefore, a hardware pipeline needs to be run economically, producing just those number of copies that are needed and no more.

While copying software has a minimal cost, steps in a software pipeline have a cost. Consider a testing step: it will require resources to perform. If the tests are regression tests and the product is complex, the tests might require a great deal of compute resources to be provisioned and used. Thus, even for software, one needs to be economical about how often one runs the pipeline; it should run when needed, and no more.

Thus, the only thing that is really different about hardware and software is the cost point. The pipeline architecture applies to both, but they each sit in a different point in terms of trade-offs between the value of running the pipeline one more time and the cost of running it. If the cost of running the pipeline is high, only run it when you have accumulated a number of product changes. If the cost is low, run it every time you make a little change.

We have already explained that one should not develop "in the pipeline." One should develop locally, where one has complete control, and use the pipeline as a final check. The question then is, what does one's "local" development setup look like?

For software, it is usually a computer—a workstation, a laptop, or a desktop running in a cloud-based virtual machine—and a set of programming tools. It might also include a virtualization system, either local or in a cloud, so that one can deploy new local versions of components (or the entire product) to run tests against them.

For hardware, a local development setup is quite similar, but instead of, or in addition to, a virtualization system, one might have tools for creating new versions of components. These tools might be a 3D printer for prototyping mechanical parts, or an electronic prototyping system for "flashing" code or circuit designs into a chip so that one can perform "hardware in the loop" tests—tests with a real hardware chip running software or implementing circuit designs that one just made changes to.

It is not that different.

A traditional way of creating a new hardware product is to have a set of teams design the hardware using digital design tools and other teams design the software. The engineers assemble these pieces, test them, collect problems, and diagnose and fix those problems. That might involve collaboration between hardware and software designers, as well as many design iterations. The goal is to deliver a working and manufacturable version by a final target date—ideally no later than the planned ship date. Traditionally, most of these steps and iterations are manual, and the only steps that are supported by automation are the design of the parts. Testing and integration are largely manual.

That approach is not a pipeline, but it works. Still, it has disadvantages. One is that it is essentially a waterfall approach, so when the first hardware and software integration is attempted, one has no idea how many problems will be found, which puts the delivery date in jeopardy. Another problem is that there is usually (thought not always) insufficient collaboration between the hardware teams and the software teams. That is a fixable problem. The team can choose to collaborate during development, but the process does not require them to do so, because the process is, by its nature, very compartmentalized.

The biggest problem is that it is a slow process and really only is viable for a product that releases infrequently. That might be okay, but increasingly today it is not.

Tesla produces hardware, specifically, cars. But it delivers substantial software updates to its cars over the air on a frequent basis. One can even order new features that are then downloaded (or simply enabled). Most auto manufacturers now use over-the-air updating, although not yet on Tesla's scale. Caterpillar updates its machines over the air. So do an increasing number of hardware manufacturers.

Of course, one cannot send new hardware over the air, but a large portion of the functionality of today's products derives from software.

Also, if you buy, say, a new phone, should you not have the latest and greatest? Why should your phone be

Agile 2 Principle: Integrate early and often.

the same as the one that your friend bought six months ago? Haven't they made any hardware improvements during the past six months? Instead of having static product versions, many companies are moving to rolling improvement, where each product that is sold has hardware improvements over those sold before, regardless of when. Tesla does that today.[2]

Multitier Products and Systems

Complex hardware products often contain components from diverse sources. For example, a vehicle always contains parts made by suppliers, as well as other parts made in house. Agile methods need to work not only for simple products but also for products that contain other embedded products, which might in turn contain smaller embedded products, like a Russian doll. Such complex multilayer products are quite common. Examples include any motor vehicle, a rocket, a computer, a refrigerator or any appliance, and a clock.

Hardware products are often divided into these levels: (1) system, (2) assembly, and (3) part. Complex products might have more levels. The "system" is the thing that is being made for customers, and it contains assemblies as well as parts. An assembly is a collection of parts (or subassemblies) that comprise some major subsystem. For example, an engine is an assembly within a car; a brake system is another assembly.

An assembly might be a product in its own right. If the brake system is made by a supplier, then it is that supplier's product. Similarly, if the brake system is made in house, then it might be treated as a product, or not. Generally, if an assembly is used in more than one other product (e.g., multiple car models), then it is best treated as an internal product; but if it is made specifically and only for use in one larger product, then it might best be treated as just a subsystem of that product rather than as its own product.

Why does it matter? By treating a subsystem as a product, one decides to generalize it so that it can be used in other products. Its lifecycle becomes decoupled from the original system in which it was used. It needs its own product manager and permanent teams. It might need its own marketing and might be treated as its own source of value instead of deriving its value only from the larger product that it was originally built for.

One challenge with complex products is that each component of the system potentially has its own development and delivery process needs. At the same time, each level of the system needs to have a development and delivery process. These all need to tie together in a harmonious way.

In Matthew Skelton and Manuel Pais's book *Team Topologies*, they explain how a static hierarchical arrangement of teams is insufficient for building a complex product. They advocate defining several different dimensions of structure, and adjusting them over time as development progresses and needs change.

They propose that most teams be organized as "stream-aligned" teams, that is, be organized around the delivery pipelines that we have described. In addition, they propose that there might need to be a "platform team," organized

> **Agile 2 Principle:** Product development is mostly a learning journey—not merely an "implementation."

> **Agile 2 Principle:** Organizational models for structure and leadership should evolve.

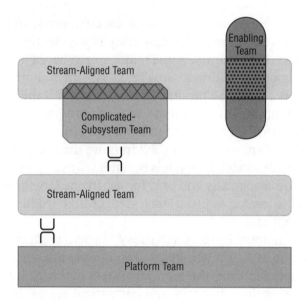

Figure 13.1: Team interaction modes

Source: Figure 7.5 of Matthew Skelton; Manuel Pais. *Team Topologies* (Kindle Location 2701). IT Revolution Press. Kindle Edition.

around any supporting infrastructure that the stream-aligned teams (pipelines) need. This is illustrated in a figure from their book, which is reproduced in Figure 13.1.

They also propose that there might need to be one or more "enabling teams" that have specialized expertise, to provide support in some manner (for which they provide several choices) to the stream-aligned teams. Finally, they propose

Agile 2 Principle: Favor mostly autonomous end-to-end delivery streams whose teams have authority to act.

Agile 2 Principle: Foster collaboration between teams through shared objectives.

Agile 2 Principle: Both specialists and generalists are valuable.

Agile 2 Principle: A team often needs more than one leader, each of a different kind.

that there might need to be some teams that focus entirely on maintaining complex subsystems. In today's cloud environments, teams dedicated to core microservices might fit the pattern of these "complex subsystem teams."

This model is quite realistic, and we have seen this pattern used many times. There is an important element that needs to be highlighted here. As we have explained in prior chapters, teams of any kind need leadership. We have described different kinds of leadership and different ways they might arise or manifest, but the bottom line is that leadership is needed. The art of setting up a team is making the important choices of what kinds of leadership are needed, who might provide those kinds or how the right people might be selected, and what kinds of authority, if any, might be needed—and then being prepared to observe and adjust over time.

Using Skelton and Pais's model, the teams that design a car would be stream-aligned. Those that design the brake system might be separate stream-aligned teams, or they might be complex subsystem teams. There might be teams that design and maintain the robotic infrastructure for building cars, and they might be a platform team. There might also be teams that provide expertise and coaching pertaining to using hardware and software design tools, and they might operate as enabling teams.

Note that these teams are not building cars; they are designing cars that can be built; and to do that, not only do they need to design the car, but they need to design the manufacturing steps needed to create the car. Thus, the stream-aligned teams might need experts in car design as well as experts in manufacturing design.

> **Agile 2 Principle:** Product design must be integrated with product implementation.

> **Agile 2 Principle:** Business leaders must understand how products and services are built and delivered.

Each of these needs leadership, but what kinds? The stream-aligned teams that design the car and how to build it need the same kinds of leadership that we have talked about for product teams. They need team leadership that can help them to get organized; they need technical leadership to help them to identify technical issues, talk them through, and get them resolved; and they need inspirational leadership to help them to work as a team.

What about a platform team? Since the platform is essentially its own product, with multiple car models sharing the same platform—that is, the same underlying robotic technologies—they might operate as their own stream-aligned team.

What about enabling teams that provide support related to the tools that the stream-aligned teams use? They might have experts in tools such as Simulink or solids modeling who can coach engineers on stream-aligned teams. These teams are there to assist and coach the development teams, not perform tasks for them or perform as a "service," which is the old approach. Once you insert another team as a "service" that performs steps for other teams, you create a queue, the development team ends up waiting, and cycle time becomes terrible.

Enabling teams need leadership. They need to operate in a helpful manner to stream-aligned teams, and they need to stay abreast of the latest tools and evaluate them. They need to operate by providing assistance rather than by performing steps, and they need leadership that inspires them; they need leadership that helps them organize; and they need leadership that markets them, in effect, to other teams and advocates for them.

What if there is a looser structure that cuts across these teams, such as a community of practice? A community of practice does not own its members. They belong to other teams, but they come together because they share a knowledge domain. Such a group still needs leadership, to help it organize, to help it stay inspired, and to advocate for it when resources are needed.

Agile 2 Principle: Technology delivery leadership must understand technology delivery.

Agile 2 Principle: Provide leadership who can both empower individuals and teams and set direction.

Agile 2 Principle: Leadership models scale.

Agile 2 Principle: The most impactful success factor is the leadership paradigm that the organization exhibits and incentivizes.

Our Case Studies

In this book there is frequent mention of both Jeff Bezos's Amazon and Elon Musk's companies Tesla and SpaceX. That is because these companies have not only changed entire industries, but because they are 21st century technology platform companies that have carved new paradigms that others can learn from. The leaders of these companies also are known for setting extremely aggressive goals.

We do not claim that these companies are the only models for success. Organizations differ. In some companies, the business strategy means more than the product strategy or the innovation strategy. During Microsoft's entire history, we are not aware of a single innovation of industry-changing significance that it can be credited with. It did not even create the operating system that launched Microsoft—it bought MS-DOS from someone else. Over the years, every big idea Microsoft had also came from someone else, but its business strategies enabled the company to triumph over competitors, driving would-be innovators like GO Corporation from the market by convincing partners to only license Microsoft's software.[3]

In his book *Extreme Teams*, Robert Bruch Shaw explains,

> *"Pixar employees believe that a great story with memorable characters is the key to making a film that people love. Story takes priority over everything else. Other studios may believe otherwise—perhaps thinking that savvy marketing or technical innovation drives a film's success."*[4]

Thus, different companies—even in the same market—can have different strategies at their core and still each be successful. As Microsoft has proven, even a tech company does not have to be innovative to be successful. In his book *Good to Great*, Jim Collins wrote,

> *"A company need not have passion for its customers (Sony didn't), or respect for the individual (Disney didn't), or quality (Wal-Mart didn't), or social responsibility (Ford didn't) in order to become enduring and great. This was one of the most paradoxical findings from Built to Last core values are essential for enduring greatness, but it doesn't seem to matter what those core values are. The point is not*

what core values you have, but that you have core values at all, that you know what they are, that you build them explicitly into the organization, and that you preserve them over time."[5]

Indeed, aggressive sales, marketing, and partnership strategies can prove very successful, even if there is little innovation going on. We do not admire that kind of success, but it is a valid kind for many.

SpaceX, Tesla, and Amazon are not just sales organizations, however: innovation is clearly a core value. Much has been written about Amazon's aggressiveness, but there can be no question about its ability to innovate. Not only has it changed the entire retail industry (not necessarily for the better), but it spearheaded cloud services, which has transformed how IT is done across the globe.

In Chapter 3 we quoted Musk saying, "I'm trying to make sure that our rate of innovation increases. . .this is really essential." He manages their rate of innovation as a strategic knob that will help him to achieve his vision.

It is not just luck either. These companies have not been one-hit wonders. Rather, they have each created winning idea after winning idea and turned those ideas into industry-changing paradigms. Clearly, there is something about how they are run that makes them relentlessly innovative. That is why we are paying so much attention to them.

Peter Drucker has said that one could either build a great company or build great ideas but not both.[6] Yet it seems that companies like Amazon, Tesla, and SpaceX show that a company *can* be both a fountain of new ideas and also be a great company. We want to understand how they do it.

Case Study: SpaceX

Let's look at SpaceX more closely.

Individuals Matter

When asked about how SpaceX and Tesla hire people, Elon Musk said that they "hire for smarts, not credentials." He went on to say,

"Would [Tesla] hire Nicola Tesla?. . .We really just look for evidence of exceptional ability. . .If Nikola Tesla applied to

Tesla, would we even give him an interview? It's not clear—
this guy came from some weird college . . . he's got some odd
mannerisms. . .it should be like, 'This kid's super smart, we'll
give him anything!'"[7]

Musk would look at an Agile or DevOps certification and no doubt dismiss it and instead would look at what someone has accomplished, in ways both related and unrelated to the work at hand, and would likely ask questions to assess how much they know and how they think.

We will discuss the difficult topic of employee evaluation in Chapter 14.

Have a Clear Path to Success

The notion of a product vision is central in Agile. However, what is missing is the notion of strategy.

An organization has a vision: its mission. Its mission is what it is trying to accomplish through its existence. But how much passion is there in the mission? Does the mission statement say something like, "We want to help people become more literate," or is it a vibrant call to action like, "We want to eliminate illiteracy in its entirety"?

According to Jack Welch,

> *"In my experience, an effective mission statement basically*
> *answers one question: How do we intend to win in this*
> *business?"*[8]

In other words, according to Welch, it does not just say what you want to do, but it says how you plan to do it. Do you plan to solve the illiteracy problem by publishing books? Or by teaching? Or by directing funding to existing organizations? In Welch's view, a mission statement should be more like, "We are going to boost SAT verbal scores by 20%." Tangible, clear-cut.

You might feel that is a strategy, not a mission. That's fair. The point, though, is to have a simple and clear-cut goal and a strategy for getting there. Call it mission or strategy; it is a guiding model for achieving the goal.

Elon Musk also has a simple way of looking at problems: (1) what forcing function will naturally push things the way that he wants them to go, and (2) what capabilities are needed to put that forcing function in place?

For example, he wants to go to Mars, so he imagined a forcing function as follows:

> *If (1) we can go for a given price per person, and if (2) there are enough people who want to go at that price, then people will go, and we will settle Mars.*

He is tackling number 1: his goal is to drastically reduce the cost of going to Mars by creating reusable and cheaply fabricated rockets using a "pipeline" approach, just as he has done for Tesla cars. For number 2, he is relying on the fact that there are already many willing people—so many that the now-defunct Mars 1 project had 200,000 volunteers.[9] The question is, could they pay the fare? Musk's goal is to reduce the price enough so that they can.

He also needs it to happen fast, because he wants a city built on Mars in his lifetime. And that is where the rate of innovation comes into play. It is also a forcing function, in that if innovation at SpaceX can become rapid, then they will get to Mars soon and solve all of the problems that are lurking to establish a self-sustaining city on Mars and that are not yet obvious.

Given that model, the details of how it is attained do not matter. What matters is that each capability is attained. Each capability has long term business value, because it contributes to the core model for success.

Agile 2 Principle: The only proof of value is a business outcome.

At this point, if you have been paying attention, you should say, "Wait! An actual outcome is getting to Mars—the capabilities developed along the way do not have individual value." And you would be right. But they get a pass, because they are part of a big bet on a disruptive change. There is no way to test some theories except by going all the way. Amazon could not tell if its strategy would work long term. One cannot test a business model that relies on the network effect or on other changes in the ecosystem. The only way for Amazon to know if its approach would work was to go the distance. Incremental success does not prove long-term success, and sometimes long term success only produces value at the end.

Musk's SpaceX model is a simple model. Thus, his vision is not just to go to Mars. His vision is really a strategy: to reduce the cost of going enough so that a natural forcing function will come into play and create an innovative culture to make it happen fast. It is tangible, and the path is clear. It gives him a way to evaluate every decision; the cost of getting to Mars must go down, and we must learn to innovate quickly; and every step along the way has to be economically viable in its own right or they will never achieve the end goal.

For the capability of getting to Mars cheaply, one can then identify sub-capabilities that are needed. One is the ability to launch a large payload to orbit—a Mars-capable rocket. This is necessary because it turns out that is the cheapest way to get people there and back. Another capability needed is the ability to create rocket fuel on Mars, because only then can the rocket not have to bring fuel with it for the return to Earth, which would drastically increase overall size, complexity, and therefore cost. The third capability is the ability to refuel in orbit, because only then can the rocket reach orbit but not have to carry its fuel for the Mars trip with it—it can launch empty but with passengers and equipment and then be fueled up in orbit.

Thus, the capability of getting to Mars cheaply has been decomposed into a series of subcapabilities. SpaceX now knows what it must learn how to do: refuel in orbit; build a rocket big enough to carry lots of people and land on Mars; and create a machine that can make and refrigerate lots of rocket fuel from the CO_2 in Mars's atmosphere and hydrogen either from water in situ or brought along. The path is clear: each subcapability is now a mission in itself and serves as the vision for each initiative.

Each subcapability can then be delegated to leaders who can provide the level of inspiration and guidance needed by their teams, in a way that leverages and motivates the best from those teams and does not stifle them and that also provides the gentle but persistent nudging pressure that is needed to move things forward relentlessly.

Having a capability model has another benefit: it tells you what metrics to strive for. Musk knows that he needs to be able to get large payloads to orbit cheaply. But how large, and how cheap?

He can tell you, because he knows the target cost per kilogram of a payload delivered to Mars to make the vision feasible. He knows the percent of payload that can be dedicated to return landing, and the percent of fuel that can be allocated for getting to orbit versus getting to Mars.

He also knows the required material tolerance required versus rocket size, because as rocket size increases, the tolerance required decreases, which increases reliability and decreases cost on a per-kilogram basis. These metrics all impact the capability model's viability for achieving the end goal.

While there are many metrics involved in the end-to-end problem of attaining Musk's long-term goal of creating a settlement on Mars, at any given point in the development of the capabilities needed to do it, there are only a few metrics that are being watched. For example, when trying to create a capability to launch a heavy payload to orbit, the metric is the cost per kilogram. When trying to land a vehicle on Mars, the metric will be kilograms that can be delivered to the surface. When trying to return a vehicle, it will be kilograms that can be returned.

Musk and SpaceX therefore have a *key metric that matters* (OMTM, which we explained in Chapter 12) for each stage of development of the Earth-to-Mars end-to-end capability.

Now you can envision what SpaceX's Mars mission business performance dashboard must look like. Further, these metrics need to be understood and internalized by everyone: they are the drivers for improvements to capabilities that decide if the mission can be achieved.

Establish Cultural Norms

An anecdote on this gentle pressure is worthwhile. Musk was concerned that at SpaceX's Boca Chica facility progress was not fast enough. Musk flies a lot internationally across time zones, so he does not have a 9–5 workday. He was able to be in Boca Chica on a Saturday evening. He scheduled a 1 a.m. all-hands, with attendance required for everyone. Imagine that: your CEO saying that you must attend an all-hands meeting in person at 1 a.m. on a Saturday evening (actually, Sunday morning).

At the meeting Musk asked, "How can we go faster?" The response was, "We need more people, and we need round-the-clock shifts." Musk then scheduled a meeting for Sunday to plan a recruiting drive, which was then held on the following Monday, and by Tuesday they had hired the required number of people—hundreds of new employees.[10]

There was no shouting at the Saturday meeting. It was Socratic. Musk simply said, "What will it take?" He wanted ideas, from anyone who was there—from the people doing the work. And then he acted as an organizer and decision-maker and took the actions needed to put his people's ideas in motion, without any delay at all. That sent some clear messages to everyone: (1) your

---

Agile 2 Principle: Good leaders are open.

---

ideas count—you just have to speak up—and (2) time matters: don't ever wait if you don't have to. Waiting is one of the seven Lean "wastes."[11]

The idea of never waiting reflects in SpaceX's approach to milestones. The company rarely announces a date by which something will be done. Instead, it announces "aspirational dates," but those dates change often. Employees try to do things as soon as they are able to. Schedules are meaningless at SpaceX: the answer to when something will happen is always "as soon as it can." Their work is not cadence-based; it is flow-based. They have no delivery dates or scheduled demos. Things happen as soon as they can possibly happen.

Holding a 1 a.m. meeting is unconventional. Few managers would even think of doing that (thank goodness). But Musk is never confined by tradition, and that has become part of the culture too. Of course, SpaceX is a little unique in that its vision is to go to Mars. That's a big goal, and it is inspiring to a lot of people. They will come to a 1 a.m. meeting for that or for something else that they believe in. But the point is that the vision needs to inspire people at some level, and it needs to guide them in their action. In this case, one part of the guidance was that *time matters*.

The other value that was reinforced by the 1 a.m. meeting, that *your ideas count*, is also exhibited by Musk's ubiquitousness in SpaceX's facilities. He comes and goes a lot, but when he is there, you see him. He is not tucked away in executive offices. He visits where the work is done, asks questions, and engages anyone and everyone in discussions. That is Gemba on steroids.

Aggressively Remove Bottlenecks

SpaceX has invested a great deal of effort in automating the process of designing and validating through simulation and delivering the machinery that they build through automation. As Elon Musk says, the company can "take the concept from your mind, translate that into a 3D object, really intuitively. . .and be able to make it real just by printing it."[12] In other words, reduce cycle time.

SpaceX uses traditional computer-aided design (CAD) tools (such as CATIA) but also invested in an end-to-end 3D modeling system to view and simulate entire assemblies and automatically print parts.[13] Importantly, the software is fast, even when handling complex assemblies, so that engineers do not have to wait, which encourages a rapid iterative design approach.

The goal is to enable rapid innovation. Musk wants engineers to be able to try new ideas as quickly as possible, even if it means making parts that will fail during testing. They call this a *hardware-rich* approach.

According to an article in Teslarati,

> "Why, then, push to launch Starship SN8 when, in Musk's own words, the probability of success is as low as '33%'?. . .SpaceX's attitude towards technology development is (unfortunately) relatively unique in the aerospace industry. While once a backbone of major parts of NASA's Apollo Program moonshot, modern aerospace companies simply do not take risks, instead choosing a systems engineering methodology and waterfall-style development approach, attempting to understand and design out every single problem to ensure success on the first try.
>
> "The result: extremely predictable, conservative solutions that take huge sums of money and time to field but yield excellent reliability and all but guarantee moderate success. SpaceX, on the other hand, borrows from early US and German rocket groups and, more recently, software companies to end up with a development approach that prioritizes efficiency, speed, and extensive testing, forever pushing the envelope and thus continually improving whatever is built."

The Agile community has another description for SpaceX's hardware-rich approach: *fail fast*. Instead of spending months or years on design and then carefully building one perfect prototype, the company builds many and tests them all in myriad ways.

For example, its new Raptor rocket engine is the most advanced rocket engine in existence, with two precombustion chambers—one fuel-rich and the other oxygen-rich. SpaceX has made a sophisticated design work, not by trying to perfect it for years before building anything and getting it just right, but by making their best design and then trying it, pushing it to its limits where it blows up, measuring everything about what happens, and then going back to the drawing board and trying again—and again and again.

SpaceX's rocket assembly line is a continuous delivery system. It is not an assembly line, because each rocket that comes out is slightly improved over the one that came before. Engineers continuously refine the steps that make the rocket, resulting in a stream of incrementally improved rockets. That is a continuous delivery pipeline.

Automation is a big part of it as well. SpaceX uses automated welding, not only because it is faster but because it results in more consistent welds. Tesla is famous for an aggressive use of robotics in its plants—so much so that Musk had to scale back his aspirations for the use of robots.[14]

> **Agile 2 Principle:** From time to time, reflect, and then enact change.

Case Study: A Major Machinery Manufacturer

This case study is kind of a counter example, because it highlights some anti-patterns that go against Agile 2 principles, and we want to show how that has an impact.

Industrial machinery is much more complex than it used to be. There was a time when a small group of engineers could design an industrial engine, machine the parts, assemble the engine, and get it to work.

Today, a commercial engine is so finely optimized that it requires many teams of people. Today's engines have finely tuned fuel injection systems that are computer-controlled; they have carefully designed combustion chambers that are designed using combustion hydrodynamic simulations. The parts are designed using 3D solids modeling systems and simulated for vibrations and resonances. Wear is simulated and parts are optimized so that they can be as light as possible but still strong where it counts. The engine has hundreds of sensors that send operating data over the Internet back to the manufacturer, where the data is analyzed to predict pending failure or the need for preventive maintenance.

Designing an engine, or any industrial machine for that matter, is no longer a matter of rules of thumb and craftsmanship. It is a digitally enabled design process, and the output is a set of digital specifications and manufacturing programs—software.

This is not to say that there is not human judgment and experience involved. There is—quite a bit, actually. For example, the company of this case study employs "engine experts" who fly in as consultants and who earn quite a bit to help to get new engine designs to work better or resolve issues, but this is because as much as it has become a science, there is still a craft element to it. Much of the craftsmanship now occurs as tweaks to numerical tables that control engine functions, and so the final result is a set of digital data.

Management Is Out-of-Date

The machinery business has a long tradition of mechanical engineering, including the hands-on crafts of machining, welding, measuring, finding unexpected sources of resonance, vibration, cracks, and wear, and generally tinkering thoughtfully to get a new machine design to work. Traditionally, designs that finally work are then converted into numerically controlled programs that automatically build the parts, which are then assembled by hand or by robots, or both, depending on the volume of production.

Things have changed, though, in waves. One wave has been the use of electronics throughout almost every aspect of a machine's operation. Another wave has been the use of digital design tools. The result is that designing an engine is not like it used to be. There is still some craftsmanship involved, as we explained, but it is now largely a digital process; and engine operation and optimization is now largely digital, even if there is craft involved in adjusting the digital parameters or design specs.

That means that engine development is now largely a software, chemistry, and physics problem. The hardware-centric rules of thumb and lessons learned from two decades ago apply only on the margin. If speed of development is a goal, what is more useful today is lessons from software engineering. The problem is, the senior management of large and established machine manufacturers do not have much experience in the domain of software engineering. They generally do not understand how to run a software organization effectively, yet that is what their company has turned into.

Management's Closed Door

Every industry changes. Change is not the problem. Change is what creates opportunity. But one must acknowledge change, and one must stay up-to-date.

In the organization that is the subject of this case study, senior management has a closed door culture. Generally speaking, someone is not permitted to talk to someone who is several levels above them. There seemed to even be an arrogance or elitism, that senior management was "above" the rank and file engineers.

In this company, one rarely sees a senior manager talking directly to engineering teams. They might talk to senior engineers who are viewed as experts, but not to the rank and file teams. They do not practice the Lean "Gemba walk" to view how work is done firsthand and ask questions.

We are not saying that they do not understand how the work is done. They generally do, but what they are missing is the ideas for improvement that the teams might have and the deeper understanding of why things are not getting better.

Instead, senior managers stay in their offices, and people come to them when invited. And since Agile and DevOps coaches

Agile 2 Principle: Good leaders are open.

are viewed as team level, they are not high enough in the hierarchy to be able to talk to the senior managers. Yet the senior management are where the problems are; it is senior management who need to learn about modern methods.

There is a challenge, however, that must be acknowledged. Agile coaches often know little about hardware, or even about programming. A lack of domain knowledge is a nonstarter for an engineering manager. An engineering manager expects that those who advise on process will have domain expertise.

We feel that is reasonable, and it is one of the things that Agile 2 seeks to address. Agile coaches need to take an interest in how the work is done. They need to acquire expertise in the latest methods, and how those can be applied in different domains, because Agile needs to look different in each domain. One cannot just apply a standard template. To help people decide what will work, one needs to understand the nature of their work. One need not be an expert, but one must know the vocabulary and have some feel for the different tasks.

At this particular company, several Agile coaches were removed after only a few months, because they tried to apply "standard" practices from experience on web application teams—practices that did not work well for the embedded software environment at this client—and were dogmatic about it, showing little interest in learning about the engineering work of the client's teams.

Remember that an Agile coach is a leader. They are not a manager; but they provide various kinds of leadership. Thus, when Agile 2 suggests that delivery leadership should understand technology delivery, Agile coaches are included in that.

Agile 2 Principle: Technology delivery leadership must understand technology delivery.

Management Does Not Understand Software

Management has a dilemma: even if it would like to learn about today's methods, who can teach it? The Agile community has not, in general, achieved standing among engineers. Engineers expect those who advise them to understand engineering to some degree. Yet today's products have become largely digital products—a perfect fit for Agile ideas. But applying those ideas requires nuance and sensitivity to the context, and one must take an interest in the domain of application.

Who can hardware engineering managers turn to? They will instinctively turn to sources within hardware engineering—not sources within software development. Thus, hardware engineering and software engineering need to synthesize their ideas. That requires each to learn about the other.

Management also needs to take it upon itself to learn and not wait to find the perfect experts. There are enough books about Agile and DevOps to do the job. As we explained, Mark Schwartz, the former CIO of the US Immigration Service—an organization with a nine digit (in dollars) annual IT spend—was not a software developer by training, but taught himself enough programming to experience it and sought books about Agile and DevOps.

These are some of the things that management needs to have a moderate level of understanding about:

- **How to design software-centric products**: Software cannot be treated as an afterthought that "makes the hardware work." The software is a product in its own right and must be treated as such.
- **How to manage software products at scale**: Software architectures can be flexible or not; they can be scalable or not; they can be robust or not. The organization needs to understand what characteristics its software has and why and match these to the product's needs. And there needs to be close bidirectional collaboration between the product vision and feature definition side and the technical vision and implementation side.
- **Dynamic environments**: This includes how dynamic environments can provide test isolation and test concurrency and why those are important.
- **Thinking in terms of a continuous flow of builds**: This is instead of phases in which a fixed body of code moves from one phase to the next.
- **Modern digital testing methods**: This includes the advantages of functional integration testing versus field integration testing and how to use test coverage to manage quality.

Management's lack of understanding of these topics is actually a common problem. Companies that grew up in an age of hardware often do not understand software. Their senior management does not know how to assess software capabilities or connect it to market needs.

In the book *Project to Product*, Mik Kersten provides the example of Nokia, which purchased the Symbian operating system in 2008. Nokia had a grand vision for leveraging Symbian and poured huge resources into an Agile transformation. Despite this and despite "doing everything right" with respect to team-level Agile practices, the effort was a complete failure. According to Kersten, the company did not address foundational product issues such as developing a technical strategy for making the operating system flexible enough to accommodate the additional features that were planned; in addition, there was a complete lack of feedback from the development side to the product planning side. Kersten wrote,

> *"I got the sense that there was no business-level under-*
> *standing of what the real bottleneck was, as the gulf bet-*
> *ween what IT and developers knew and what the business*
> *assumed was so vast."*[15]

and,

> *"Nokia had an engine and infrastructure for innovating on*
> *the hardware front that was a pinnacle of the Age of Mass*
> *Production, but they did not have an effective engine and*
> *infrastructure for the Age of Software and did not have the*
> *management metrics or practices in place to realize that until*
> *it was too late."*

Management knew how to develop innovative hardware products, but it did not understand how to develop software products, and it did not know what it did not know.

Management Does Not Apply Systems Thinking

In the company in question, we noticed an absence of systems thinking with respect to how the product development teams work. This was surprising since systems thinking is a core aspect of electrical engineering training, and there are electrical engineers in management at this firm; but the firm does tend to be dominated by mechanical engineers. Mechanical engineering espouses systems thinking, but it is not as critical. In contrast, systems thinking is essential in electrical engineering, because electronic systems are typically designed to have an enumerable number of states (otherwise there is a race condition); mechanical systems are not designed from that perspective.

We observed that management seemed to have little interest in understanding product development as a system. We often heard that if software has issues, it is the engineer's "fault"—that the engineers should not design products that have bugs. This is old-school thinking based on a work paradigm in which each person practices a craft, designing and fashioning a part; but in a complex digital system, there are often unexpected interactions between parts, so one must view it as a system and view the development process as a system that systematically identifies and removes those unexpected interactions. We never heard the problems framed in a system-wide context—only in localized contexts.

From time to time we did see efforts by management to make things work better, but again, these changes seemed to originate from high above, with little discussion that included those who would be affected. The changes also tended to be about organization structure, rather than focused on learning, leadership, or practices that affect how the work is actually done or how risk is managed. Thus, they did not seem to be driven by end-to-end goals or a system-oriented understanding of end-to-end efficacy.

Quality Managed Mostly Through Field Testing

Management did not appear to understand the advantages of functional integration testing versus field integration testing or how to use test coverage to manage quality. This was apparent in the lack of comprehensive test suites for the products. The result was that products were completed with lots of issues that were only discovered through field testing. Quality was managed through months of use of equipment by operators in field tests. If there was a problem, it would surface then.

That is a workable strategy, but it forgoes the rapid turnaround of early tests. Finding defects through field testing means that when a problem occurs, one has to send a "dump" of the product's internal state to an engineer and hope they can figure out what happened. That is an inefficient process, and if the problem is found, then the engineer will make a fix, but the engineer cannot fully validate the fix, because there is no comprehensive test suite, so the fix needs to be redeployed to the field, with lots more field testing done. Often a fix creates additional problems. Thus, while quality can be managed through this process, product feature lead time is

Agile 2 Principle: Integrate early and often.

extremely long, and product fixes have less assurance that there will not be unexpected side effects.

In this setting, a fix tends to be treated as if one is patching a hole, but each fix changes the system and can make the engine unstable. There is no "regression suite" to validate a fix.

Teams Pulled in Multiple Directions

The product teams supported their own products, which is a key DevOps practice. But in the cases we saw, there was no differentiation between who was providing support one week or working on new product features. Multiple business area managers would contact a team, each asking for an urgent fix or change requested by a large customer. This disrupted the flow of work and caused much angst because teams could not keep their commitments, since the team members had been spending time on "urgent" customer requests. There was no coordinated filtering or prioritization of the work heaped on teams. There actually was an attempt to address that at one point, so perhaps the situation has improved.

The core problem was that no one was setting direction. Should a team work on tasks in its backlog, or should the team members drop everything to handle every incoming business manager request? Or does it depend, and if so, who decides? It was being left to the team members or team lead to decide, but they were still being held accountable for their progress against their backlog.

The challenge was that development managers did not want to push back against the powerful business area managers, to say "We don't have time to work on your request this week." Business area managers have connections at the upper levels of product delivery management. Such informal relationships are championed by Agilists, but it can be problematic if someone goes around the proper channels to exert influence and teams or managers at lower levels become scapegoats when things do not get done.

Thus, informal communication is powerful if it is positive, but it can be destructive if it is used in self-serving ways; informal communication is an opportunity for misuse of one's influence.

There is no perfect solution for this dilemma. An organization relies on the good intentions of its members. If people have self-serving intentions, then no structure or set of rules will solve that. That is why it is essential to create a culture that promotes and gives power to those who have positive intentions, rather than those who tend to be self-serving. Leadership is everything; having the right leaders is therefore also everything.

Agile 2 Principle: Provide leadership who can both empower individuals and teams and set direction.

Each Software Tier Is a Silo

This company's engine products have several layers of software. One layer is a real-time operating system (custom designed) that runs on custom chips and interacts with the various sensors and solenoids of the engine. The next layer up is a kind of control and reporting layer that tells the engine what to do and reports on its status. The top layer is an application layer that has aggregate functions for various things such as fuel operation and emission control. The application layer also sends data to a networked telematics electronic control unit (ECU), which in turns sends data back to data center servers for monitoring and analysis.

The lowest layer of the software is generalized so that the chips and their operating system can be used in many different engines. However, it is often the case that when new sensors or sensing modes are added to an engine, this lowest layer needs to be changed, or the "codes" that it sends change. That impacts the higher layers. One might assume that this has been abstracted out so that such changes would not impact the middle layer, but in fact changes to the middle layer are often needed, since they aggregate and interpret the signals coming from the lowest layer. In software terms, slight refactoring of the entire stack is often needed.

The problem is that those who maintain the lowest layer are embedded software engineers, and they code in C. The middle and application layers are maintained by application programmers, and they use higher-level languages such as C++. There is a de facto gulf between the lowest-layer teams and the upper-layer teams.

The lowest layer tends to be viewed as a "platform" such that upper layers rest on that platform. However, given that the lowest layer tends to change a little but on a somewhat frequent basis, and sometimes change a lot on a less frequent basis, we wonder if it should not be treated as a complex component, rather than a platform. A platform should be highly stable, but the lowest layer here does not seem to be. Perhaps it cannot be.

If viewed as a component rather than a platform, the elite status of the lowest layer would be normalized, and it would be treated as a peer of the higher layers. Breaking changes to either would justify refactoring of the entire system. Right now, deference is always given to the lowest layer teams, because they are the "platform" and have a higher status socially within the organization—because they are the "real" engineers who code directly to hardware, rather than mere programmers working far above the hardware.

We feel that engineering elitism has led to the silos that exist: the three layers of the ECU software architecture. However, the elitism might be justified. The engineers will want to know if those who maintain the upper layers truly appreciate the nuances and challenges of the lowest layer and the hardware on which it runs. If they do not, then the engineers have a point.

Such intellectual silos always need to be challenged. Sometimes there are true experts. There are some skills that take years to learn and that require a substantial foundation to even begin to learn them. Medicine is like that. So is machine learning, with respect to building custom neural network models. But in the case of this company, is the hardware level of software—the operating system—really so far beyond the programmers who work on the upper layers? Or could some of those who work on upper layers, who volunteer and want to invest the effort, learn enough about the lowest level to be able to have intelligent discussions about it? If so, then it means that there could be a more collaborative relationship, and occasional refactoring of the entire stack would be more feasible.

Notes

1 en.wikipedia.org/wiki/Ransom_E._Olds

2 www.teslareporter.com/1552/take-a-look-at-all-the-continuous-changes-made-to-teslas-over-the-years

3 caselaw.findlaw.com/us-4th-circuit/1480859.html

4 Robert Bruce Shaw. *Extreme Teams: Why Pixar, Netflix, Airbnb, and Other Cutting-Edge Companies Succeed Where Most Fail*. AMACOM. Kindle Edition, p. 144.

5 Jim Collins. *Good to Great*. Harper Business. Kindle Edition, pp. 176–177.

6 Greg Mckeown. *Essentialism*. Crown. Kindle Edition, p. 55.

7 www.youtube.com/watch?v=Opnk-cPOM50&app=desktop

8 Jack Welch, Suzy Welch. *Winning*. HarperCollins e-books. Kindle Edition, p. 14.

9 www.space.com/22758-mars-colony-volunteers-mars-one.html

10 arstechnica.com/science/2020/03/inside-elon-musks-plan-to-build-one-starship-a-week-and-settle-mars

11 en.wikipedia.org/wiki/The_Toyota_Way

12 www.youtube.com/watch?v=xNqs_S-zEBY

13 www.geoplm.com/knowledge-base-resources/GEOPLM-Siemens-PLM-NX-SpaceX-cs-Z10.pdf

14 economictimes.indiatimes.com/magazines/panache/elon-musk-calls-excessive-use-of-robots-at-tesla-factory-mistake-says-humans-are-underrated/articleshow/63777620.cms

15 Mik Kersten. *Project to Product*. IT Revolution Press. Kindle Edition, p. 64.

14

Agile 2 in Service Domains

Agile 2 ideas are not just for technology products. The whole point of Agile 2 is to help humans work together to accomplish things—all kinds of things! Indeed, we have seen people use Agile ideas and techniques for planning chores around the house, for improving surgical procedures, and for building construction. Pretty much any human endeavor involving more than one person can potentially benefit from looking at it with Agile 2 ideas in mind.

This does not mean that every Agile 2 idea applies to every situation, and it does not mean that any Agile framework can be used as a starting point for any activity; rather, it means that there are Agile 2 ideas (and other ideas too) that might help.

There is one nontechnical function that is present in nearly every product and service organization: human resources (HR). In this chapter, we will consider the impact of Agile 2 on human resources and discuss a new HR model known as PeopleOps, one of several modern reinterpretations of HR from a "people and culture" perspective.

We will also provide some examples of the use of Agile 2 ideas in a nonproduct and highly human-centric domain: healthcare clinical delivery. Our intention is not to show how Agile 2 ideas can be used in a whole range of non-product activities: that would justify many books just about that. This chapter merely attempts to illustrate that Agile 2 is not just for building products and technical stuff.

In preparation for the discussion, let's consider the role of HR in organizations. Traditional HR functions treat people as resources. All too often an HR function actually exists to benefit the organization —to deal with problems that people might raise and make sure that the organization complies with employee-related laws and is not vulnerable to employee lawsuits. As such, HR is not an advocate for the employee. Employees know this, and it means that there is no one in the organization who the employee can turn to in order to discuss their situation in a trustworthy way.

That is actually a loss for all involved, since employees who are dissatisfied will not be dedicated, and if you have learned one thing from this book, we hope it is the importance of having people be enthusiastic about their work and inspired by their organization.

The PeopleOps movement[1] provides a new way of looking at HR. The central idea behind PeopleOps is that HR staff should be treated as partners who help to achieve business goals. PeopleOps is somewhat nascent at this writing and its defining Agile PeopleOps Framework (APF) is still being developed, but we will share those aspects that have been defined that are reflected in Agile 2.

In the APF, a PeopleOps team consists of those in the organization whose focus is on the *people* of the organization. Today we would call this team the HR department, but as we have explained, the scope and approach of PeopleOps are quite different from how HR is usually practiced. Thus, it is not a mere renaming.

The healthcare industry has long embraced Lean methods. We have already referenced Dr. Atul Gawande's book *The Checklist Manifesto*. We can suggest some additional Agile 2 ideas that are sometimes applicable in healthcare settings.

Define a Target Culture

Edgar Schein, the father of organizational culture, has said, "The only thing of real importance that leaders do is to create and manage culture. If you do not manage culture, it manages you, and you may not even be aware of the extent to which this is happening."[2]

Building a healthy and positive workplace culture should not be treated as a one-time activity after post-culture assessments. Instead, leaders need to intentionally practice a culture of positive behaviors and reinforce them on a regular basis so that these behaviors become a habit over a period of time.

> **Agile 2 Principle:** The most impactful success factor is the leadership paradigm that the organization exhibits and incentivizes.

We have explained how and why leadership needs to establish a culture of the right kinds of leadership and the right behaviors. According to the APF, a PeopleOps team provides leadership with research-based information on organizational culture and helps them understand the critical roles of culture in an organization. This serves as a guiding start to reflect and ideate on the kind of culture to be built within the organization.

The PeopleOps team then acts as a collaborative partner to help define the elements of culture that are important for realizing the organization's goals. Creating the right culture becomes an important strategy in support of other business strategies.

Recall from our discussion in earlier chapters that there are different kinds of leadership: helping to organize and coordinate things; helping to reflect, discuss, and decide; helping to act; helping to ideate; being accountable. Leadership can include authority for making decisions, but that is not what defines leadership. Recall also that leadership can be outward focused, toward stakeholders, constituents, or the public; or it can be inward focused, toward one's team or the members of one's organization.

Some organizations specifically choose a strategy of pitting employees against each other, under the theory that the most aggressive and competitive ones will serve the organization best. That might be true in some organizations, but it is rarely true in a product development context, where the work is highly creative, a lot of collaboration is needed, and technology implementation has a strategic impact on success. In a product development organization, cultural elements that foster cooperation are the most powerful and lead to the most product agility. However, in a sales organization, it is common to have people compete for bonuses based on their sales numbers.

We have also seen organizations that take Jack Welch's strategy about weeding out the "bottom 10%" to heart and each year cull a fraction of people who are the lowest performers by some measure. Netflix uses this approach. According to Robert Shaw, the author of *Extreme Teams*,

> *"Netflix emphasizes that people and teams are fully accountable for the results they produce. If people prove unworthy of the freedom they are provided, based on a lack of performance, they are given a generous severance package. If someone stumbles and fails to perform, he or she is given time to recover—but not too much time."*[3]

It is not up to us to say what the best strategy for an organization should be. What we can say with certainty is that if you

Agile 2 Principle: A team often needs more than one leader, each of a different kind.

leave culture as an afterthought, you will not have control over your organization, and sustained success will be unlikely.

Culture consists of behaviors, prevalent knowledge, and attitudes. An Agile organization needs to have the following cultural norms:

- **Forms of leadership that are fair and that encourage**— These should be accepted and seen as normal, with toxic forms of leadership immediately recognized and shunned.
- **Thoughtful (not reckless) experimentation**
- **Measurement as a standard practice**—for all important outcomes and behaviors
- **Continuous transparency**—Measures are published for all to see, throughout an initiative, without fear; and criticism is not tolerated, but thoughtful critique with suggestions for improvement are viewed positively and seen as normal and not competitive. This is not an easy thing to achieve in many organizations!
- **Working incrementally but still end-to-end**—as the normal way of doing things
- **Thinking end-to-end**—instead of in a narrow this-initiative-only context
- **Open door**—Anyone can talk to anyone, and that is not threatening or inappropriate.
- **Socratic discussion**—collaboration and discussing ideas on the merits. This should be a normal practice, in which all embrace finding the best answer, rather than the idea that they proposed.
- **Empowerment, accountability, and integrity**—People are expected to take the lead to decide how to do the things that they are asked to do; they are expected to have integrity on outcomes, with transparency, and accepting of helpful critique and suggestions; they are expected to be accountable for the outcome; but managers are also expected to not "punish" for unfavorable outcomes, but instead to have a constructive approach.
- **Outcomes are not all that matters**—Outcomes are really important, but behavior matters too, because behavior becomes

an organizational norm, and toxic behavior leads to a toxic organization. An organization that only rewards good outcomes is a toxic and competitive place where people do not cooperate, every opportunity is viewed in a zero-sum manner, and politics becomes the dominant form of decision-making.

A PeopleOps function can help to create strategies to grow these norms within the organization's culture. Inevitably, this must include demonstration of these norms by leadership, to set an example. It will usually also include training and advocacy of these norms, and alignment of incentives. *All* members of the organization must receive this training.

Regarding the first bullet, "Forms of leadership that are fair and that encourage," imagine if the HR or PeopleOps function is *actually trusted* by employees. If trusted, it could survey the staff about leadership and ask which leaders are effective and fair. This is significant, because toxic leaders are often able to cultivate a strong relationship upward, creating the impression that all is well.

However, as Jack Welch once said, "CEOs can talk and blab all day about culture, but the employees know who the jerks are."[4] Imagine if HR could provide accurate feedback about how well the organization's managers are actually doing in leading their people.

> **Agile 2 Principle:** Provide leadership who can both empower individuals and teams, and set direction.

> **Agile 2 Principle:** Good leaders are open.

> **Agile 2 Principle:** Validate ideas through small, contained experiments.

Managers will need additional training in how to recognize these norms and how to reward and reinforce them. The drive to establish these norms must be persistent, spanning years; and senior leadership needs to embrace these norms itself, because people will immediately observe if there is a double standard, where leadership is not practicing the norms that they are advocating for.

We refer again to the Transtheoretical Model (TTM), which defines stages of change that an individual goes through. An organization goes through those stages as well, in an aggregate sense. First people must see that the organization wants things to change, and people must become convinced that it is not a "fad." That will happen only if they see senior leadership demonstrating the new norms and if this persists. Of course, there are those who will immediately embrace change, but most people will take a wait-and-see approach.

If people become convinced that the intention to change is for real—that management really means it and intends to change their own behavior—then people will accept the change as legitimate and start to visualize how their own behavior might change. This is where situational training can help. Simulation of situations, in which people have to make decisions, and their decisions can be influenced by old norms or by the new norms.

Over time, if management persists in advocating for the new norms, then as people start to change their own behavior, the new norms will start to become established and seen as "how we do things." If the pressure to change is sustained, then it will become rooted and will become the new norm for real. Years will have passed. As Jim Collins wrote in his seminal book *Good to Great*,

> *"Sustainable transformations follow a predictable pattern of buildup and breakthrough. Like pushing on a giant, heavy flywheel, it takes a lot of effort to get the thing moving at all, but with persistent pushing in a consistent direction over a long period of time, the flywheel builds momentum, eventually hitting a point of breakthrough."*[5]

Recognizing the Right Leaders

Leaders are people who others follow for any reason, by definition. Picking the right leaders is therefore essential for changing an organization's culture. People look to their leaders for cues of what behavior is desired as well as what is tolerated.

It is unfortunately the case that often bad behavior is hidden from view. Harassment, for example, often occurs during one-on-one interactions, out of sight of those who might intervene. It is really important for people to not only be prepared to recognize good leadership and good behavior but also to recognize bad behavior—so-called anti-patterns.

Some kinds of bad behavior include the following:

- **Taking over**—telling others what to do, rather than suggesting; or becoming an organizer but not allowing anyone else to also organize things. An example is ignoring or speaking dismissively of others' attempts to organize activities.
- **Forming an inner circle**—creating confidences among a subset of a group. This is particularly toxic if certain members are discussed in a competitive way, creating cross purposes within the team.
- **Not giving credit to others**—expressing the ideas of others without mentioning that the idea came from those others
- **Not being transparent or owning up to errors or lapses**—conveniently leaving out mention of one's errors, or actively concealing them, to manage "optics"
- **Overpowering others in discussions or in any forum**—speaking loudly, or using body language to overpower others; or speaking so continuously or forcefully, without leaving pauses so that others can speak. Other examples are not asking those who are silent what they think, and not proactively seeking the thoughts of others who are less vocal or less aggressive.
- **Focus on winning instead of finding the correct answer or approach**—dismissing the ideas of others if those ideas might prove superior to one's own ideas. Arguing as if in a debate, where the objective is to win, instead of the objective being to find the truth or find the best idea.

Through training, people can learn to recognize these patterns of behavior. Not only do the cultural norms therefore need to include certain behaviors, but norms also need to include rejection of toxic behaviors that are counter to how the organization wants to operate, consistent with its strategies.

Anonymous surveys are helpful to assess how authoritative leaders are viewed by those who they lead. It is far too easy for someone in a position of authority to cultivate a false view of their leadership. Surveys, if truly anonymous, will reveal a pattern of whether a leader's followers actually respect their leader.

The risk is that management will hear of negative feedback and seek to identify and purge the complainers. One of us had a friendly People and Culture team in an agency he worked for. The People and Culture team sent a quarterly attitude survey to staff that included questions about what needed improvement. The questions also asked what area people worked in. Whenever there was any negative feedback about leaders, the CEO and COO would get angry, try to identify who said it, and then target them for criticism, ostracism, and elimination. They did this only based on suspicions.

That kind of behavior is reminiscent of a fascist or criminal organization that stifles dissent and ruthlessly punishes all those who are not loyal to the leader. It is not conducive to the exchange of ideas, and it is antithetical to the climate that is needed for the creation of innovative products.

Create Incentives to Encourage the Desired Behaviors

Senior management must demonstrate the desired behaviors themselves if they expect others to adopt those behaviors. However, demonstrating behavior is not enough. One must also provide training—ideally situationally experiential—and one must also create reinforcing activities and align incentives with the desired behaviors.

Consider a healthcare organization that we interviewed for this book. That organization has adopted Lean methods to improve efficacy. To track adoption and progress, the organization created a scorecard. The top half of the scorecard tracked use of desired behaviors, and the bottom half tracked outcomes.

They use a system of metrics for the scorecard. These include metrics pertaining to efficacy, consisting of five sets of measures. Clinicians can add additional metrics. They also made metrics part of treatment, instead of an add-on, so that the metrics are updated in the course of treatment, instead of after the fact.

In addition to instituting behavioral and efficacy metrics, they also perform assessments, including the following:

- Patient surveys
    - Audits: which practices are being used?
- Drills to test competency
- Patient chart reviews
- Q&A in which clinicians can ask questions and discuss issues with Lean advocates

This institution learned that when measuring outcomes that are subjective, they need to use language carefully. For example, when asking patients in a survey a question such as, "Do you feel safe in your home?" the patient might misunderstand the intention or scope of the question. Subjective survey questions need to be extremely clear and explicit about what they are asking. They also need to take into account that some people might feel insecure about answering honestly, particularly if they do not trust the intentions of those who are asking.

That is why it is essential for a PeopleOps function to earn the trust of personnel, as being not just an instrument for management, but as a team who has the individual's interests at heart as well and can be trusted with confidentiality and impartiality. For this to be possible, the organization's leaders must behave in an open and inquisitive way, wanting to hear the truth, rather than behave like fascist leaders who punish anyone who does not exhibit unquestioning loyalty. If your leaders are of the latter kind, your organization cannot become an Agile organization.

PeopleOps demands that organizational leaders and HR representatives are transparent in their approach; that they do as they say they will, with integrity; and that they can be trusted.

About scorecards, what about the HR or PeopleOps function? How well is it running? And is the effectiveness and development of people measured in any way in the various teams that people belong to? In *DevOps for the Modern Enterprise*, Mirco Hering wrote,

> *"almost every time I ask directors what they use to measure the quality of people management in their teams; often, they have no measures whatsoever."*[6]

We measure how well we use computing infrastructure. We measure how well we use office space. People are the most important resource, yet we often do not systematically measure how well they are utilized, not just in the short term, but in the long term. We are not talking about time utilization: we are talking about talent utilization.

Leaders as Coaches and Mentors

People need coaching and training for their work skills and for leadership, including being able to work on a self-organizing team.

The PeopleOps philosophy advocates that managers and other leaders engage in coaching conversations with their team members. This is a core responsibility of a manager, and their performance is assessed partly on the degree to which they coach and mentor their organization's or team's members.

At the healthcare organization that we have described, it was found that behavioral health clinicians had a strong preference for paper (a habit) and were not accustomed to using an electronic chart. They would wait to receive a lab report by fax, even though the data was in the online chart. They would then review the report and then write up their own report, print it, and give it to an administrative assistant to scan.

They also were accustomed to paper reports in which each report contained all of the relevant sections, in text, and were unaccustomed to an electronic chart in which the information was present in other tabs or sections of the chart and so it did not need to be replicated as text in the clinician's own report. The clinicians would actually copy sections from other parts of the chart and paste them into their report so that it looked like a traditional paper report.

The solution was for their clinical supervisors to coach them individually on using an electronic chart and the new norms associated with that. Mere training did not suffice; they required follow-up and someone to keep reminding them and showing them what they had done. These were experienced but nontechnical people, and they had ingrained habits in terms of how they conducted their work. Those habits needed to be changed.

Process Varies with Circumstances and Time

Things change over time. A strategy that made sense when a company was a startup might not make sense any longer when it has grown to be 30 people and then 100 people. The strategy might need to change again when it grows to have thousands of employees. Circumstances can also change, requiring a change in strategy.

The previously mentioned healthcare organization had been using online patient questionnaires so that patients could fill in the information ahead of time instead of taking time during an office visit. This worked well until the COVID-19 pandemic occurred. During the pandemic, office visits became online visits, so once the patient was online, it became more efficient to capture the information right then, instead of having them separately fill in online forms ahead of time. It was a flow examination that revealed that this new approach would be more efficient.

Agile 2 Principle: Validate ideas through small, contained experiments.

Agile 2 Principle: Organizational models for structure and leadership should evolve.

Reexamination needs to occur at all levels. Have circumstances changed assumptions that underlie current strategies? The availability and cost of staff and skills is a continuing issue in technology platform companies; in fact, the situation changes almost monthly. The world is also becoming more globally connected. Attitudes about how people want to work change over time. HR departments need to track how attitudes change.

The market changes rapidly too: one year MySpace was the most popular social media platform, the next year Facebook was. Constant re-evaluation is necessary to stay ahead.

This means the biggest threats include the information silos that exist in so many organizations. If an organization's leaders do not have an accurate view of what is happening in their organization and what is happening in the market, they will make bad choices and bad decisions. A culture of openness and safe expression of ideas is the only way to ensure that leadership is receiving enough points of view that they can learn if they are not considering something important.

The Dysfunction of Staffing Functions

In the prior chapter we mentioned that Elon Musk once reflected, "Would we even give Nikola Tesla an interview?" His point was that organizations often pay too much attention to someone's certificates and other academic achievements, because those often have little relationship to how one performs in the real world. HR departments and recruiters need to accept that there is no formulaic approach to finding the right people on behalf of technology partners.

Indeed, there is enormous dysfunction today in how technology professionals are recruited. So many organizations still treat technology professionals as commodities, even though those individuals are building the organization's make-or-break strategic business platform. If you are building rockets, don't you want a rocket scientist? The same applies to any technology if your business hinges on it: you need the very best who you can get.

Low-cost suppliers of "resources" do not have those people. People who are highly talented do not work through low-cost commodity vendors. They often join or found boutique firms, or they become the most senior technical staff at elite companies. This means that if your organization is unwilling to do business with a boutique vendor, then those people are out of reach; and when a truly talented person joins a large consulting firm, they will be too valuable to use by placing them on-site; instead, they will be used to help to close deals. In general, the best people are available from boutique firms, so when an organization will not do business with a boutique firm, they forego the ability to find the best and the brightest.

Staffing vendors that recruit for technology often employ people who know little about the technology. Those of us in technology professions are accustomed to getting calls from staffing vendors who go down a list of "buzzwords," asking the silliest question of all: "How many years of experience do you have with XYZ?"

It is a silly question because today most technologists do not use some technology tool—some "XYZ"—day after day for some period of years. It does not work like that. What usually happens is that we use "XYZ" here and there for a while, and then move on to some other initiative and might not touch XYZ for several years. So when a recruiter asks, "How many years of experience do you have with XYZ?" the question does not make sense. That is not how technologists work anymore.

Asking about specific tools is also an extremely poor way to identify the right people. Imagine that a software developer is an expert in Microsoft Azure cloud architecture services. Now imagine that they receive a call and the recruiter asks, "How many years of experience do you have with Amazon Web Services?" The developer will say "none," and the recruiter will say "Thank you" and hang up. However, the services that Amazon has are highly similar to the services that Azure has. If someone is expert with one, they have a level of maturity that conveys easily to the other cloud system. The patterns of judgment are the same, and those patterns of judgment are what are most valuable;

the recruiter completely missed that, because they did not know how to discuss these services with the developer.

Recruiters should be looking for people who have the aptitude to learn new things and leverage knowledge they have to solve similar types of problems and those who have intellectual curiosity.

Agile 2 Principle: Professional development of individuals is essential.

Technologists are accustomed to learning new tools all the time. It is part of the job. What is valuable is their experience with a range of tool types and kinds of product architectures. Knowledge of specific tools is no longer telling about how effective someone will be. But recruiters for commodity staffing companies do not know how to recruit someone in any way other than going down a checklist of tool names and certifications, which they generally know little about.

Agile 2 Principle: Different Agile certifications have unequal value and require scrutiny.

What is far more important than specific skills, and even smarts, might be one's level of immersion in the domain in which the organization operates. Does the person "get" what the organization does? Do they embrace that as something important? In his book *Extreme Teams*, Robert Shaw recounts how the company Patagonia realized that the most effective employees are those who relate to the company's business. According to Shaw,

> *"As the company [Patagonia] grew, it increasingly looked for people with more traditional business backgrounds and specialized skills (what the founder calls the MBA types). After a number of these people failed in their roles, the firm went back to its roots. . .Those who work for Whole Foods need to love natural food. Those who work for Pixar need to love storytelling."*[7]

Google learned that it is foolish to recruit based on specific skills. In their book *Lean Enterprise*, Humble, Molesky, and O'Reilly write,

> *"Google's recruiting strategy is liberating because it enormously expands the pool of qualified applicants. Instead of*

looking for 'purple squirrels' with precisely the skills and experience required for a job, we should look for people who can rapidly acquire the necessary skills and then invest in an environment that enables them to do so."[8]

Connection to the mission is important for managers too, and indeed all leaders in an organization. The well-known organizational psychologist Dr. Marta Wilson speaks to this in her book *Leaders in Motion*. She writes,

"By engaging the heart of everyone in important transformational goals, leaders prepare groups to share accomplishment, which would not be possible if all or most of them were not motivated by a heart connection to the vision . . .Until change occurs in the hearts around us, the workplace cannot transform, and we cannot achieve authentic or sustainable progress."[9]

If one should not hire for specific tools and skills, how do people obtain those skills? Organizations that want to have a creative environment *need* to invest in their people by providing training and by giving them time and resources to learn, while setting the expectation that learning is necessary.

Yes, *some will leave* when they acquire new skills, but if you have a positive environment, others will come to take their place. People who are creative and natural learners do not want to work in a place where they are not able to learn new things.

Also, only training employees and not contractors makes no sense anymore, since in knowledge-intensive organizations, the work that contractors perform is usually not well-defined tasks but includes all of the multifaceted things that employees do, and one cannot find all those things on a contractor's résumé. Who pays for it is another issue, but both employees and contractors need the same training and learning opportunities if they are going to work on the same things.

Another challenge is that when an organization asks a commodity staffing firm to obtain people, they will usually specify a bracketed compensation range. The recruiters call people and say "What is your rate?" or "What is your salary expectation?" and if you state something that is out of their range, they hang up. That filters out the best and the brightest, because the best and brightest are often just beyond the bracket, which is usually set aggressively low; or a highly effective person might be significantly beyond the bracket.

Imagine losing someone who is ten times as effective as the average employee—a true change agent and evangelist for new methods and a wellspring of new ideas—and losing them because they were 10 percent (or even 50 percent) beyond your compensation bracket.

Balance Short- and Long-Term Views of People

There is a tension between having people work on things that provide immediate value and developing those same people so that they might provide more value, or other kinds of value, in the future. It is often the case that project leads are focused mostly short term, so it falls to others in an organization to advocate for the long-term development of an employee.

If people report only to an initiative, project, or functional team, it is natural for them to be viewed by that group's leadership entirely in that context. It will be difficult for a typical manager to want to enable their people to "look elsewhere" or try other things within the organization, or learn skills that would only be useful in other parts of the organization.

A PeopleOps function or a resource manager—someone who can advocate for the long-term development of the employee in the larger context of the entire organization—needs to mentor the person to make sure they are not "pigeonholed" so that they are developed for the benefit of the organization as a whole, and not just their current function. Basically, there needs to be someone with whom the employee can sit down to map out a set of goals for the next year or two, and take those into account when selecting which teams the person might be put on. If there is no such person and they do not have enough influence to get the person on those teams, then the employee's growth will be stunted.

The benefit also needs to be for the employee, because otherwise, they will view what benefits them as separate from what benefits the organization. An individual's professional development strategy needs to represent a "win-win" for both the individual and the organization.

Professional development needs to be embedded in an organization's culture; otherwise, it will be difficult to advocate for an employee with respect to their development, because professional development is always an investment, and the benefit is always beyond the immediate. It therefore represents an immediate short-term cost.

The Agile PeopleOps Framework (APF) emphasizes the professional development of organizational members from day one of their journey with the organization.

The APF also introduces what it calls *personas*. These are not intended to prescribe roles or job titles; they are illustrative. The persona "manager" serves as a mentor for the team members. That implies that the manager can see beyond the current team's purpose, and place value on the development of the individual team members, to benefit the organization but apart from the team. For a manager to have that perspective, there needs to be a strong cultural value placed on the development of people within the organization.

The APF stresses the need to maximize the potential of people, instead of keeping them "in their lane" where they perform the same function optimally but eventually burn

---

Agile 2 Principle: Individuals matter just as the team matters.

Agile 2 Principle: Professional development of individuals is essential.

---

out or become bored and move on. Of course, for this to work, PeopleOps needs to be driven from the executive level.

Managers and leaders need to have conversations on a regular basis to discuss how career development can be shaped for the organization's members. The development of people needs to be an organization-wide strategy, shaped by the kinds of leadership and the kinds of skills and abilities that will be needed as other strategies unfold.

Investing in the professional development of organizational members is directly proportional to the productivity and high performance of teams, provided the individual development plans and careers paths are charted with the employees.

Design an Effective Work Environment

It is a great paradox that today some of the most highly paid nonmanagement staff are often given less floor space, quiet, and privacy than is given to an administrative assistant who is paid far less. This is not to begrudge administrative staff; it is to point out that nonmanagement staff are subject to a "penny-wise but pound foolish" approach to their work environment.

During the 1980s it was standard practice to give engineers, which included programmers, a private office. During the 1990s that changed, and the new norm became a cubicle. Today the norm is to squeeze a large number of software engineers onto a long table or put them in "pods" in an "Agile team room" configuration. Yet that arrangement, as we have explained, is not conducive to collaboration, which was the rationale for the setup, and is really terrible for focus. According to a paper in the *Journal of Organization Design,*

> *"Unplanned interruptions are among the most effective productivity killers for knowledge workers. Working in an office environment with constant distractions makes it difficult for many knowledge workers to achieve high productivity, and practically impossible to engage in so-called deep work."*[10]

Instead of copying a template for how teams should be set up in terms of their floor space and work environment, *ask them.* Perform a survey. Think through how they will spend their day. How often do they need to talk with others? How does this cohort of people communicate best and most naturally? What do they spend most of their time doing, and what setup is best for that?

Discuss the issue with professionals whose specialty is workplace productivity. Avoid dogmatic people who have an agenda or a particular approach that they are trying to perpetuate. Prefer to ask experts who take a neutral approach, considering each situation uniquely, rather than a one-size-fits-all approach.

Above all, if you pay people a lot, don't penny pinch on their work environment! What a waste of money to hire expensive people and then handicap them by making them work in an office arrangement that makes them inefficient or ineffective.

Agile 2 Principle: Respect cognitive flow.

Agile 2 Principle: Make it easy for people to engage in uninterrupted, focused work.

Assess Performance Immediately

Every manager dreads having to do annual performance reviews. Guess why? Because the review is so far removed from any specific instance of an employee's performance that it becomes an evaluation of the person—personal.

There is also the reality that today much of the work that people do on teams changes throughout the year. As a result, an annual review would need to assess each situation spanning the many changes in a person's activities.

Regardless, annual reviews are useless. They are *too late*. They cherry-pick specific instances that the manager happens to recall and ignore all the rest of a person's behavior all year long. Worse, managers often have no clue about how someone actually works with their peers—not unless the manager is a hands-on leader who talks to everyone on the team every day, which is what they *should* be doing, but all too often do not.

People need immediate, contextual feedback. If they do something great, they should be told right away. If they do something not so great, they should be told right away—and given a chance to explain it from their perspective. Do not let the matter languish. Usually the best person to give the feedback is the leader or leaders who are most directly involved with that aspect of the person's work such as the tech lead, team lead, or other leadership role.

Immediate feedback creates a short feedback loop, which enables someone to improve right away. It also maintains morale because when someone's performance is not good, they might be aware of it, and if they have discussed it and agreed on a strategy for addressing it, they are back "in good" with their leaders. It creates a positive atmosphere of continuous improvement.

The organization's HR or PeopleOps function needs to define performance feedback practices that are effective. These practices need to be defined as organizational norms, and managers need to be trained in their use.

Practices should include immediate feedback in the form of coaching, as well as recurring and frequent check-ins between managers and their staff as well as staff and their career mentors. This provides the employee with an ongoing path of interaction between them and their manager and mentor during their professional development journey.

Companies like Deloitte, Netflix, and Adobe are working toward building a frequent check-in culture. According to Robert Shaw, writing in *Extreme Teams*,

> *"[Netflix eliminated] performance reviews within the firm, which were replaced with ongoing supervisor/employee discussions combined with periodic peer feedback that asks what each person should stop, start, or continue."*[11]

Submarine commander David Marquet echoes the need for feedback to be immediate in his book *Turn the Ship Around*. He even states it as a fundamental principle, hence the capitalization:

> *"USE IMMEDIATE RECOGNITION TO REINFORCE DESIRED BEHAVIORS is a mechanism for CLARITY."*[12]

Feedback also should be specific rather than general. In *DevOps for the Modern Enterprise*, Mirco Hering writes,

> *"Your feedback should focus on concrete situations that are recent, not generalizations such as 'you are often late.'"*[13]

If feedback fails to be about specific incidents, the person cannot visualize the situations in question, and the feedback then becomes about them as a person instead of about specific behaviors. If the feedback recalls specific incidents, then they can reflect on those to consider how they might have behaved differently in those situations and alter their behavior in the future. If they are able to visualize how they might behave differently, they have a plan, and the discussion about it has been successful and was a win-win between them and their manager or supervisor. The outcome is positive instead of negative.

Notes

1 agilepeopleopsframework.com
2 www.ifcc.org/media/478037/sedef-yenice-organizational
   -culture-and-managing-change-pathcape.pdf
3 Robert Bruce Shaw. *Extreme Teams: Why Pixar, Netflix, Airbnb, and Other Cutting-Edge Companies Succeed Where Most Fail.* AMACOM. Kindle Edition, p. 160.

4 Jez Humble, Joanne Molesky, Barry O'Reilly. *Lean Enterprise*. O'Reilly Media. Kindle Edition, Kindle Locations 4228–4229.

5 Jim Collins. *Good to Great*. Harper Business. Kindle Edition, p. 186.

6 Mirco Hering. *DevOps for the Modern Enterprise*. IT Revolution Press. Kindle Edition, p. 194.

7 Robert Bruce Shaw. *Extreme Teams: Why Pixar, Netflix, Airbnb, and Other Cutting-Edge Companies Succeed Where Most Fail*. AMACOM. Kindle Edition, p. 110.

8 Jez Humble, Joanne Molesky, Barry O'Reilly. *Lean Enterprise*. O'Reilly Media. Kindle Edition, Kindle Locations 4492–4495.

9 Dr. Marta Wilson. *Leaders in Motion: Winning the Race for Organizational Health, Wealth, and Creative Power*. Greenleaf Book Group Press. Kindle Edition, p. 117.

10 jorgdesign.springeropen.com/articles/10.1186/s41469 -017-0014-1

11 Robert Bruce Shaw. *Extreme Teams: Why Pixar, Netflix, Airbnb, and Other Cutting-Edge Companies Succeed Where Most Fail*. AMACOM. Kindle Edition, p. 158.

12 L. David Marquet. *Turn the Ship Around!*. Penguin Publishing Group. Kindle Edition, p. 187.

13 Mirco Hering. *DevOps for the Modern Enterprise*. IT Revolution Press. Kindle Edition, p. 189.

15

Conclusion

Agile 2 brings many additional ideas into the fold of Agile. These ideas are often used by those who practice Agile well, but they have not been part of Agile per se—yet they need to be, because Agile cannot work well without them.

Some important ideas and topics are misunderstood or misrepresented by much of the Agile community. One of them is leadership. The importance of leadership is well established: people have written about it for thousands of years. In the context of product development, some important recent books provide ample discussion of leadership and its importance. For example, in *Accelerate*, by Forsgren, Humble, and Kim, there is an entire chapter on leadership. In that chapter, they ask,

> *"Why have technology practitioners continuously sought to improve the approach to software development and deployment as well as the stability and security of infrastructure and platforms, yet, in large part, have overlooked (or are unclear about) the way to lead, manage, and sustain these endeavors?. . .we must improve the way we lead and manage IT."*[1]

Mark Schwartz, wrote in his book *A Seat at the Table*,

> *"Agile approaches seem to remove IT leaders from the value-creation process. In an Agile process, visioning, refinement, and acceptance of system capabilities are in the hands of product experts and users—that is, the folks from the business side. Delivery teams work directly with users and product owners from the enterprise lines of business to decide what is valuable and to create solutions.*

Where does this leave IT leaders who have always believed themselves responsible for making sure that IT delivers business value?"[2]

We hope that from this book you have learned why leadership is crucial for human collective endeavors, and that we have provided some useful models for different kinds of leadership and when they can be effective, while avoiding prescribing what kinds of leadership to use. We hope that we have provided ideas for how to reimagine your organization's leadership culture and how to bring that about and make it sustainable.

Some other areas that have not been explicitly part of Agile per se, but need to be, are product design and data. These are hugely important topics that need to be addressed holistically and in an Agile manner, in consideration of the entire product envisioning, delivery, and value creation context.

A Model for Behavioral Change

The discussion of the Transtheoretical Model (TTM) was included to provide a framework for thinking about change. Too often change is dealt with in an ad hoc way, either with mostly an emphasis on providing inspiration, or changing culture, or through demonstrating change and a tolerance for experimentation. Those are all essential; but in the course of an organization's change, each person goes through their own change lifecycle, and the TTM is useful for considering how people change their thinking and their approach to things over time, because organization change cannot happen unless individuals change.

The Agile community has been too dogmatic and its interpretation of Agile ideas too extreme and narrow, with a one-size-fits-all mindset. This behavior has limited Agile's effectiveness and contributed to an enduring distrust of Agile among much of the engineering community and parts of the business and senior management community. Whether it is the forcing of everyone into an open "team room" or insisting that all collaboration begin face-to-face, Agile culture is rife with practices that are, too often, forced on people without taking into account that people are not homogeneous. Change is unattainable if the change will not work for a large percent of the organization's members.

We hope that Agile 2 has provided a bridge to engineering through Agile 2 statements about the importance of technical concerns throughout the value stream, and through principles related to technical leadership, integration, and the value of specialization when it is called for. We feel that specialists should not be treated as deviations from a generalist ideal, but instead that both specialists and generalists are seen as valuable and that the challenge is to stitch their work together so that it is continuous.

The Agile community has also been too insular, seeming to pretend that resistance to some of its extreme methods, such as open team rooms, does not exist (search the Internet for *open plan office* to see the many articles against the practice) and ignoring pleas from introverts (just search the Internet for *Agile and introverts*) to take them into account. The community has also long downplayed the importance of technical development issues, even though Agilists like Jez Humble presented important technical ideas at Agile conferences. The Agile community at large's failure to embrace these ideas forced them into a separate domain (DevOps). We feel that DevOps should not be apart from Agile but that the two should merge back into a whole.

No More Tribalism

Finally, the community has fractured into tribes. There is the Scrum tribe (which itself has split), the Kanban tribe, the SAFe tribe, the LeSS tribe, the DevOps tribe, a few smaller tribes, and then there is everyone else—those who do not want to belong to a tribe. They are the ones with a true Agile mindset.

These dysfunctions need to be fixed. Otherwise, Agile will become the problem, rather than the solution. Agile will become a legacy paradigm, replaced by something else (DevOps?).

Agile 2 attempts to make these adjustments, because we feel that Agile is important. The core ideas are highly important—if they are taken in a nonextreme manner and nuance is added to them.

We need people to view Agile 2 not as a new dogma but as an attempt to fine-tune many Agile ideas and to explicitly open Agile to the broader set of ideas contributed by other domains. We are calling to Agilists (and we include the DevOps community) to reject alignment with any particular Agile "tribe" and instead be inclusive. This does not mean the intellectual contributions of those tribes are not valuable; they are. But none of us should not confine ourself to a single tribe.

Agile 2 explicitly asks its readers to not be insular. The message is loud and clear. If you are reading mostly books that have "Agile" in their title or "Scrum" or some other Agile buzzword, then you are not reading broadly enough; and if you view a particular Agile tribe as your main source of insights, then you also are not thinking broadly and inclusively enough.

Agile Cannot Be Simplified

Agile 2 does not replace Agile. Rather, it attempts to modify it and add to it. Agile 2 embraces without duplicating many other domains of thought, including Lean, DevOps, and any others that might apply in a given situation. Agile 2 does not attempt to be "feature complete" in any sense and does not replace anything, including the Agile Manifesto. Agile 2 is a set of ideas that we felt were not getting enough attention or that were being viewed in the wrong way.

However, we reject attempts to distill Agile into simpler forms. There is no "essence" of Agile: any attempt to distill Agile will inevitably be misleading in many situations. Issues pertaining to human behavior are not simple. Short maxims can often inspire people, and that is a good thing; but short maxims do not contain answers. The Agile movement and its Manifesto provided inspiration; Agile 2 is about helping people to find answers.

Even Agile 2's principles are not absolutes. No principle pertaining to human behavior can ever be. Agile 2's principles are guidelines to consider—not rules to follow.

Agile Is Timeless

Agile 2's principles will likely be revised slightly, but Agile 2 will not be followed by Agile 3, Agile 4 and so on. *We believe that these ideas need to be solid and timeless*, not ever-evolving.

We do not believe there is such a thing as "modern" Agile (or "more modern" Agile, or "*even* more modern Agile," and so on). Agile ideas may be refined and added to, but corrections imply mistakes! We are sorry, but that is true, because Agile is about human behavior, and humans are not evolving—at least not fast enough to make a difference. That is why each generation repeats the same mistakes.

Culture changes over time, and youth culture is always different from what came before. People today seem to have less patience: young people check their phone continually and have been trained by computer games to expect immediate gratification and constant action. The ability of people to focus seems to be waning, although that is no surprise given the environments they are expected to work in.

But are people really different? Human DNA does not change appreciably over one generation, but people's behavior is also determined by changes that occur during their early development. According to Dr. Jim Taylor, writing in *Psychology Today*,

". . . [childrens'] brains are still developing and malleable, frequent exposure by so-called digital natives to technology is actually wiring the brain in ways very different than in previous generations."[3]

Thus, the changes are not only cultural, but might also be affecting us at a biological level.

Perhaps the changes are not all bad. Perhaps younger people are able to scan more information quickly, having been exposed to so much of it. Yet, we have not observed that. If anything, people are more resistant to reading today than ever. It seems they have less patience for anything that is in-depth and expect information to be no longer than a Twitter tweet. That is *not* a good change, and we should not accept it, because the effect of it is to dumb us down and reduce the complexity of dialogue and therefore what we can achieve.

We also must embrace cultural and technology changes that make new work arrangements possible. We now can collaborate globally, but to do that, we need to work differently than before. We can contact each other instantly and around the clock, but we need to define boundaries around that, to preserve our ability to focus as well as our work-life balance.

Perhaps if we create the right work environment and are cognizant of cognitive load theory and stop piling Agile ceremonies and distractions onto people, we can adapt and restore our ability to focus and "go deep." Perhaps we need to ease up on the time pressure of the relentless sprint cadence and focus more on outcomes, which might give people more mental breathing room. Perhaps we can bring focus and depth back to product development. Indeed, if one looks at organizations such as Amazon and SpaceX, they clearly have been able to accomplish game-changing things, and they are led by quiet reflective thinkers who place a high value on deep dialogue and long-form writing—bucking today's trends.

We hear that computer games have created a culture of instant gratification, but we have not noticed that developers need instant gratification. They know that complex products take time.

Perhaps the changes that have been claimed are superficial. Perhaps people have not changed that much on the inside and the brains of even mature humans are still very plastic. Maybe not much has actually changed that cannot be undone.

Just as when someone moves from one culture to another, they can adapt; so we believe that if the right norms are established, people can learn to work in a healthy and effective manner, regardless of what youth culture is doing. People still change a great deal after they have left youth behind.

We want Agile to be cross-cultural, not reliant on particular cultural norms or values; we want Agile to reflect what humans are capable of, not the dumbed-down expectations of today's society.

Ideas for how humans can work together have not changed over time. Humans were as imaginative during the time of the Great Pyramids as they are today. What Thucydides wrote about human behavior in groups is as valid today as it was then. Socratic leadership is still as useful today as it was in his time, and there were servant leaders in ancient times even though they did not go by that name.

Agile thinkers did not invent new leadership or management styles—they just raised awareness of them. That is why Verne Harhish's book *Mastering the Rockefeller Habits* reads like an Agile playbook, and it is why Kenneth Roman, the former CEO of Ogilvy & Mather Worldwide, the famous ad agency, describes his former boss and company founder's habits and they also sound like they came from an Agile playbook.[4] The idea of a small, self-organizing team most likely dates to prehistoric times.

The only irrefutable differences today are that the rate of change has increased, and technology is increasingly important. The same humans exist in an ever-changing technological substrate of our own creation. Humans have not changed in any permanent way (yet), and so the essence of human dynamics has not changed.

We hope that Agile 2 helps you in your human endeavors to find ways to work best with others and to accomplish your greatest goals!

Notes

1 Nicole Forsgren PhD, Jez Humble, Gene Kim. *Accelerate*. IT Revolution Press. Kindle Edition, pp. 238–239.
2 Mark Schwartz. *A Seat at the Table*. IT Revolution Press. Kindle Edition.
3 www.psychologytoday.com/us/blog/the-power-prime/201212/how-technology-is-changing-the-way-children-think-and-focus
4 marker.medium.com/lessons-from-a-legendary-advertising-agency-about-being-a-great-boss-8bf054c74870

Index